Basic Numerical Mathematics

ISNM

INTERNATIONAL SERIES OF NUMERICAL MATHEMATICS
INTERNATIONALE SCHRIFTENREIHE ZUR NUMERISCHEN MATHEMATIK
SÉRIE INTERNATIONALE D'ANALYSE NUMÉRIQUE

VOL 22

Basic Numerical Mathematics

Vol. 2: Numerical Algebra

by

John Todd
Professor of Mathematics
California Institute of Technology

ACADEMIC PRESS New York San Francisco London 1978
A Subsidiary of Harcourt Brace Jovanovich, Publishers

BIRKHÄUSER VERLAG BASEL UND STUTTGART 1977

Licensed edition for North and South America, Academic Press, Inc.,
New York/San Francisco/London
A Subsidiary of Harcourt Brace Jovanovich, Publishers.

ACADEMIC PRESS, INC.
111 Fifth Avenue, New York, New York 10003

LIBRARY OF CONGRESS CATALOG CARD NUMBER:

ISBN 0-12-692402-3

PRINTED IN THE UNITED STATES OF AMERICA

Contents

Notations and Abbreviations

$\delta(i, j)$ or δ_{ij} is defined by: $\delta(i, j) = 0$ if $i \neq j$, $\delta(i, i) = 1$.

$\mathbf{R}_n$ (resp. $\mathbf{C}_n$) denotes the space of *column* vectors with n real (resp. complex) components.

Vectors will be denoted by lower case black type so that, e.g., if $\boldsymbol{a} \in \mathbf{R}_n$ then

$$\boldsymbol{a} = \begin{bmatrix} a_1 \\ \vdots \\ a_n \end{bmatrix}$$

where the a's are real.

$\boldsymbol{e}_i$ will denote the ith unit column vector, all components being zero except the ith, which is 1 so that $[e_i]_j = \delta(i, j)$.

$\boldsymbol{e}$ will denote the column vector all of whose components are 1.

Matrices will be denoted by upper case black type so that, e.g.,

$$\mathbf{A} = \begin{bmatrix} a_{11} & a_{12} & \dots & a_{1n} \\ \vdots & \vdots & & \vdots \\ a_{m1} & a_{m2} & \dots & a_{mn} \end{bmatrix} = [a_{ij}]$$

is an $m \times n$ matrix. We shall usually dispense with commas between the elements a_{ij} and between their subscripts.

$\mathbf{I}$ or $\mathbf{I}_n$ will denote the unit $n \times n$ matrix: $[\mathbf{I}]_{ij} = \delta(i, j)$.

$\mathbf{J}$ will denote the matrix all of whose elements are 1.

If $a_{ij} = 0$ for $i \neq j$ we shall write $\mathbf{A} = \operatorname{diag}[a_{11}, \dots, a_{nn}]$.

Transposition of matrices (including vectors) will be denoted by a prime or dash so that $\mathbf{A}'$ will be an $n \times m$ matrix and $\boldsymbol{a}'$ a row vector. We will use * to denote *conjugate transposition* so that, e.g.,

$$[\mathbf{A}^*]_{ij} = \bar{a}_{ji} \quad \text{and} \quad \boldsymbol{a}^* = [\bar{a}_1, \dots, \bar{a}_n].$$

We shall occasionally write row_i to indicate the ith row of a matrix and similarly use col_j.

The *inner* (or *scalar*) *product* of two vectors will be denoted by

$$(\boldsymbol{x}, \boldsymbol{y}) = \boldsymbol{x}'\boldsymbol{y} = \sum x_i y_i.$$

The *determinant* of a square matrix $\mathbf{A}$ will be denoted by $\det \mathbf{A}$.

The *trace* of a square matrix $\mathbf{A}$ is $\operatorname{tr} \mathbf{A} = \sum a_{ii}$.

The *characteristic polynomial* of a square matrix $\boldsymbol{A}$ is $\det(\boldsymbol{A}-x\boldsymbol{I})$. The *characteristic values* of $\boldsymbol{A}$ are the zeros of its characteristic polynomial and are usually denoted by $\alpha_1, \ldots, \alpha_n$ and it is often assumed that $|\alpha_1| \geqq |\alpha_2| \geqq \ldots \geqq |\alpha_n|$. In the latter case α_1 is called the *dominant characteristic value* and $|\alpha_1|$ the *spectral radius* of $\boldsymbol{A}$.

A *characteristic vector* of $\boldsymbol{A}$ corresponding to a characteristic value α is a (non-zero) solution of the equation $\boldsymbol{A}\boldsymbol{a}=\alpha\boldsymbol{a}$. We then speak of $(\alpha, \boldsymbol{a})$ as a *characteristic pair* of $\boldsymbol{A}$.

The *pseudo-inverse* of a (rectangular) matrix $\boldsymbol{A}$ will be denoted by $\boldsymbol{A}^I$: when $\boldsymbol{A}$ is square and $\det \boldsymbol{A} \neq 0$, $\boldsymbol{A}^I = \boldsymbol{A}^{-1}$.

Norms of vectors and matrices will be denoted by $\|\boldsymbol{a}\|$ or $\|\boldsymbol{A}\|$ with appropriate subscripts.

We use the standard *order* symbolism, e.g., $f(n)=\mathcal{O}(g(n))$ as $n\to\infty$ means that there exists n_0, A such that when $n \geqq n_0$ (depending in general on A) we have $|f(n)| \leqq A|g(n)|$. Often $g(n)$ will be a power of n.

$\sum'$ will be used to indicate (or emphasize) the omission of a special term in a sum, e.g.,

$$\sum{}' |a_{ij}| = \sum_{\substack{j=1 \\ j\neq i}}^{n} |a_{ij}|.$$

We use $\doteqdot$ to indicate approximate equality.

We use the standard logical symbolism e.g., $\Rightarrow$ for implies, $\in$ for belongs to, $\notin$ for does not belong to, $\subset$ for is included in.

Preface

There is no doubt nowadays that numerical mathematics is an essential component of any educational program. It is probably more efficient to present such material after a reasonable competence in (at least) linear algebra and calculus has already been attained — but at this stage those not specializing in numerical mathematics are often interested in getting more deeply into their chosen field than in developing skills for later use. An alternative approach is to incorporate the numerical aspects of linear algebra and calculus as these subjects are being developed. Long experience has persuaded us that a third attack on this problem is the best and this is developed in the present two volumes, which are, however, easily adaptable to other circumstances.

The approach we prefer is to treat the numerical aspects separately, but after some theoretical background. This is often desirable because of the shortage of persons qualified to present the combined approach and also because the numerical approach provides an often welcome change which, however, in addition, can lead to better appreciation of the fundamental concepts. For instance, in a 6-quarter course in Calculus and Linear Algebra, the material in Volume 1 can be handled in the third quarter and that in Volume 2 in the fifth or sixth quarter.

The two volumes are independent and can be used in either order — the second requires a little more background in programming since the machine problems involve the use of arrays (vectors and matrices) while the first is mostly concerned with scalar computation.

In the first of these, subtitled "Numerical Analysis", we assume that the fundamental ideas of calculus of one variable have been absorbed: in particular, the ideas of convergence and continuity. We then take off with a study of "rate of convergence" and follow this with accounts of "acceleration process" and of "asymptotic series" — these permit illumination and consolidation of earlier concepts. After this we return to the more traditional topics of interpolation, quadrature and differential equations.

Throughout both volumes we emphasize the idea of "controlled computational experiments": we try to check our programs and get some idea of errors by using them on problems of which we already know the solution — such experiments can in some way replace the error analyses which are not appropriate in beginning courses. We also try to exhibit "bad examples" which show some of the difficulties which are present in our subject and which can curb reckless use of equipment. In the Appendix we have included some relatively unfamiliar parts of the theory of Bessel functions which are used in the construction of some of our examples.

In the second volume, subtitled "Numerical Algebra", we assume that the fundamental ideas of linear algebra: vector space, basis, matrix, determinant, characteristic values and vectors, have been absorbed. We use repeatedly the existence of an orthogonal matrix which diagonalizes a real symmetric matrix; we make considerable use of partitioned or block matrices, but we need the Jordan normal form only incidentally. After an initial chapter on the manipulation of vectors and matrices we study norms, especially induced norms. Then the *direct* solution of the inversion problem is taken up, first in the context of theoretical arithmetic (i.e., when round-off is disregarded) and then in the context of practical computation. Various methods of handling the characteristic value problems are then discussed. Next, several *iterative* methods for the solution of system of linear equations are examined. It is then feasible to discuss two applications: the first, the solution of a two-point boundary value problem, and the second, that of least squares curve fitting. This volume concludes with an account of the singular value decomposition and pseudo-inverses.

Here, as in Volume 1, the ideas of "controlled computational experiments" and "bad examples" are emphasized. There is, however, one marked difference between the two volumes. In the first, on the whole, the machine problems are to be done entirely by the students; in the second, they are expected to use the subroutines provided by the computing system — it is too much to expect a beginner to write efficient matrix programs; instead we encourage him to compare and evaluate the various library programs to which he has access.

The problems have been collected in connection with courses given over a period of almost 30 years beginning at King's College, London, in 1946 when only a few desk machines were available. Since then such machines as SEAC, various models of UNIVAC, Burroughs, and IBM equipment and, most recently, PDP 10, have been used in conjunction with the courses which have been given at New York University, and at the California Institute of Technology.

We recommend the use of systems with "remote consoles" because, for instance, on the one hand, the instantaneous detection of clerical slips and, on the other, the sequential observation of convergents is especially valuable to beginners. The programming language used is immaterial. However, most of the problems in Volume 1 can be dealt with using simple programmable hand calculators but many of these in Volume 2 require the more sophisticated hand calculators (i.e. those with replaceable programs).

The machine problems have been chosen so that a beginning can be made with very little programming knowledge, and competence in the use of the various facilities available can be developed as the course proceeds. In view of the variety of computing systems available, it is not possible to deal with this aspect of the course explicitly — this has to be handled having regard to local conditions.

We have not considered it necessary to give the machine programs required in the solution of the problems: the programs are almost always trivial and when they are not, the use of library subroutines is intended. A typical problem later in Volume 2 will require, e.g., the generation of a special matrix, a call to the library for a subroutine to operate on the matrix and then a program to evaluate the error in the alleged solution provided by the machine.

Courses such as this cannot be taught properly, no matter how expert the teaching assistants are, unless the instructor has genuine practical experience in the use of computers and a minimum requirement for this is that he should have done a significant proportion of the problems himself.

Basic Numerical Mathematics

CHAPTER 1

Manipulation of Vectors and Matrices

The objective of this chapter is to familiarize those who have been accustomed to scalar computation with computations involving arrays. In the problems, in general, it is sufficient to take $n=5$, to avoid excessive print out. The matrices $\mathbf{A}_1, \mathbf{A}_3, \ldots$ in Problem 1.2 are taken from a larger list of test matrices; it is suggested that the reader chooses one of these matrices and uses this throughout whenever a choice has to be made.

REMARK: We later use $\mathbf{A}_1$ or $\mathbf{A}_{11}$ to denote *special* matrices (or submatrices) related to a general matrix $\mathbf{A}$. Reference to the context will resolve any ambiguities.

Chapter 1. Problems

1.1. Print out the n-vector:

$$\boldsymbol{e}'=[1, 1, \ldots, 1].$$

Print out the n-vector $\boldsymbol{v}_i$, where the j-th component of $\boldsymbol{v}_i$ is

$$(\boldsymbol{v}_i)_j=(2/(n+1))^{1/2} \sin ij\pi/(n+1),$$

say, for $i=2$.

Print out the n unit vectors $\boldsymbol{e}_i$, where the j-th component of $\boldsymbol{e}_i$ is δ_{ij}.

1.2. Print out the $n\times n$ unit matrix, $\mathbf{I}$.

Print out the $n\times n$ matrix $\mathbf{A}=[a_{ij}]$ where

(i) $\mathbf{A}_1: a_{ij}=(2/(n+1))^{1/2} \sin (ij\pi/(n+1))$.
(ii) $\mathbf{A}_3: a_{ii}=n+1, \quad a_{ij}=1 \quad \text{for} \quad j\neq i$.
(iii) $\mathbf{A}_7: a_{ij}=i/j \quad \text{if} \quad i\leqq j, \quad a_{ij}=j/i \quad \text{if} \quad i>j$.
(iv) $\mathbf{A}_8: a_{ii}=-2, \quad a_{ij}=1 \quad \text{if} \quad |i-j|=1, \quad a_{ij}=0 \quad \text{if} \quad |i-j|\geqq 2$.
(v) $\mathbf{A}_9: a_{ij}=2 \min (i,j)-1$.
(vi) $\mathbf{A}_{11}: a_{ij}=|i-j|$.
(vii) $\mathbf{A}_{15}: a_{jk}=\exp 2\pi i(j-1)(k-1)/n$.
(viii) $\mathbf{A}_{13}: a_{ij}=(i+j-1)^{-1}$, the Hilbert matrix $\mathbf{H}_n$.
(ix) $\mathbf{A}_{16}: a_{ij}=\binom{i}{j-1}, \quad j=1, 2, \ldots, i-1, i; \; a_{ij}=0, \; j>1.$

1.3. Write a program to calculate $a\mathbf{A}$ where a is a constant.

1.4. Write a program to calculate $\mathbf{A}+\mathbf{B}$ where $\mathbf{A}$ and $\mathbf{B}$ are two matrices of the same size.

1.5. We define $|\mathbf{A}|=[|a_{ij}|]$. Write a program to calculate $|\mathbf{A}|$. Apply it to some of the above cases.

1.6. Write a program to calculate $\mathbf{C}=\mathbf{AB}$ when the product is defined, i.e., if $\mathbf{A}$ is ${}_n\overset{m}{\boxed{}}$ and $\mathbf{B}$ is ${}_m\underset{p}{\boxed{}}$ then $\mathbf{C}=\mathbf{AB}$ is ${}_n\underset{p}{\boxed{}}$.

Verify that this works in extreme cases, e.g., the case of a scalar (or inner) product $x'y$ and the case of an anti-scalar product xy', where x, y are column vectors. The last case turns up often in numerical analysis.

1.7. Write a program to compute $\sum_i \sum_j a_{ij}$ where the a_{ij} are the elements of a matrix $\mathbf{A}$.

Evaluate these sums for the case when $\mathbf{A}$ is the *inverse* of the matrices $\mathbf{A}_3$, $\mathbf{A}_{13}$ for some small values of n. From these results, conjecture the value of these sums in general. Establish your conjecture in the case of the matrix $\mathbf{A}_3$.

1.8. Print out the scalar products of pairs of rows and pairs of columns of the matrix $\mathbf{A}_1$.

1.9. Let $\mathbf{A}$ be an $n\times n$ symmetric matrix. The expression

$$x'\mathbf{A}x=(\mathbf{A}x, x)$$

is a scalar, the quadratic form

$$\sum_i \sum_j a_{ij}\, x_i\, x_j.$$

In many problems in matrix theory (cf. Chapter 10), the Rayleigh quotient $R(x)$, associated with a vector $x\neq 0$ and a symmetric matrix $\mathbf{A}$, and defined by

$$R(x)=\frac{x'\mathbf{A}x}{x'x}$$

is of great importance.

Write a program to calculate $R(x)$, given $\mathbf{A}$ and x. Use it to calculate $R(x)$ in some special cases, e.g., when $\mathbf{A}=\mathbf{A}_8$ and when

$$x=w_i \quad \text{where} \quad (w_i)_j=\sin ij\pi/(n+1), \quad j=1, 2, \ldots, n.$$

If $\alpha_1 \geqq \alpha_2 \geqq \ldots \geqq \alpha_n$ are the characteristic values of $\mathbf{A}$ show that

$$\alpha_1=\max_{x\neq 0} R(x), \quad \alpha_n=\min_{x\neq 0} R(x)$$

1.10. If w is a vector of unit length so that $w'w=1$ show that $\mathbf{0}=\mathbf{I}-2ww'$ is orthogonal, i.e., $\mathbf{0}\mathbf{0}'=\mathbf{I}$.

Write a program to compute $\mathbf{0}$ given w and then use the multiplication program to compute $\mathbf{0}\mathbf{0}'-\mathbf{I}=\mathbf{A}$.

Find $\max |a_{ij}|$ and $\sum_i \sum_j |a_{ij}|$ for some vectors w.

1.11. For α real we define $\operatorname{sign} \alpha=0$ if $\alpha=0$, $\operatorname{sign} \alpha=1$ if $\alpha>0$, $\operatorname{sign} \alpha=-1$ if $\alpha<0$.

For a real vector $x=[x_1, x_2, \ldots, x_n]'$ define $\delta(x)$ to be the number of changes of sign in the sequence

$$\operatorname{sign} x_1, \ \operatorname{sign} x_2, \ \ldots, \ \operatorname{sign} x_n.$$

Write a program to compute $\delta(x)$ and check it on some cases.

CHAPTER 2

Norms of Vectors and Matrices

Much work done on computers with vectors and matrices is approximation mathematics, and it is necessary to be able to say when one vector is near another, or when a vector is small, and similarly for matrices. For this purpose the idea of norm was introduced. In most cases the norm of a 1-dimensional vector or matrix is the absolute value of the number.

We begin with three vector norms in common use:

(1) euclidean norm: $\|\boldsymbol{x}\|_2=[\sum|x_i|^2]^{1/2}$.

(2) maximum, Chebyshev or sup norm: $\|\boldsymbol{x}\|_\infty=\max|x_i|$.

(3) Manhattan norm: $\|\boldsymbol{x}\|_1=\sum|x_i|$.

These three norms, which are non-negative functions on the n-dimensional vector space $\mathbf{R}_n$ or $\mathbf{C}_n$, satisfy the following vector norm axioms:

(V1) $\|\boldsymbol{x}\|\geqq 0$ and $\|\boldsymbol{x}\|=0$ if and only if $\boldsymbol{x}=0$.

(V2) $\|\alpha\boldsymbol{x}\|=|\alpha|\,\|\boldsymbol{x}\|$.

(V3) $\|\boldsymbol{x}+\boldsymbol{y}\|\leqq\|\boldsymbol{x}\|+\|\boldsymbol{y}\|$.

A consequence of (V3) is

$$\|\boldsymbol{x}-\boldsymbol{y}\|\leqq|\,\|\boldsymbol{x}\|-\|\boldsymbol{y}\|\,|.$$

Direct proofs can be given: alternatively, the results come as special or limiting case of Problem 2.4. Of these axioms (V2) expresses strict homogeneity and (V3) can be interpreted as the triangle inequality.

We next discuss some matrix norms:

(4) Schur or Frobenius norm $\|\mathbf{A}\|_F=[\sum\sum|a_{ij}|^2]^{1/2}$.

(5) max absolute row sum $\|\mathbf{A}\|_\infty=\max\limits_i\sum\limits_j|a_{ij}|$.

(6) max absolute column sum $\|\mathbf{A}\|_1=\max\limits_j\sum\limits_i|a_{ij}|$.

(7) $\|\mathbf{A}\|_M=n\max\limits_{i,j}|a_{ij}|$.

(8) $\|\mathbf{A}\|_2=$[dominant characteristic value of $\mathbf{A}'\mathbf{A}]^{1/2}$.

It can be verified that all of these satisfy the matrix norm axioms.

(M1) $\|\mathbf{A}\| \geqq 0$, $\|\mathbf{A}\| = 0$ if and only if $\mathbf{A} = 0$.

(M2) $\|a\mathbf{A}\| = |a| \, \|\mathbf{A}\|$.

(M3) $\|\mathbf{A}+\mathbf{B}\| \leqq \|\mathbf{A}\| + \|\mathbf{B}\|$.

(M4) $\|\mathbf{AB}\| \leqq \|\mathbf{A}\| \, \|\mathbf{B}\|$.

These results can be established directly in all the above cases; alternatively the results for the cases (5), (6) and (8) follow from Theorem 3.1.

Observe that the "norm"

(7′) $$\|\mathbf{A}\| = \max_{i,j} |a_{ij}|$$

does not satisfy the submultiplicative axiom (M4). In fact, if

$$\mathbf{A} = \mathbf{B} = \begin{bmatrix} 1 & 1 \\ 0 & 1 \end{bmatrix} \text{ then } \mathbf{AB} = \begin{bmatrix} 1 & 2 \\ 0 & 1 \end{bmatrix}$$

and

$$\|\mathbf{AB}\| = 2, \quad \|\mathbf{A}\| = \|\mathbf{B}\| = 1.$$

Chapter 2. Problems

2.1. Write a program to calculate each of the three vector norms and use it to compute the norms of the vectors of Problem 1.1.

2.2. Write a program to calculate all the three norms of a vector by one pass along it.

2.3. Evaluate some of the above matrix norms for some of the matrices.

2.4. Show that

$$\|\mathbf{x}\|_p = \sqrt[p]{\sum |x_i|^p}$$

is a norm in $\mathbf{R}_n$ when $p \geqq 1$. Is this so for $0 < p < 1$?

It is clear that this p-norm has the euclidean and Manhattan norm as special cases (indicated by the notations in the text). Show that the Chebyshev norm is a limiting case of the p-norm

$$\lim_{p\to\infty} \|\mathbf{x}\|_p = \max |x_i| = \|\mathbf{x}\|_\infty.$$

2.5. Indicate the sets of points (or two-dimensional vectors) such that

$$\|\mathbf{x}\|_1 \leqq 1, \quad \|\mathbf{x}\|_2 \leqq 1, \quad \|\mathbf{x}\|_\infty \leqq 1.$$

What properties do these sets have in common?

2.6. Sketch the curve $x^{2/3}+y^{2/3}=1$. If we write, for two-dimensional real vectors $\boldsymbol{x}=[x_1, x_2]'$,

$$\|\boldsymbol{x}\|=[|x_1|^{2/3}+|x_2|^{2/3}]^{3/2}$$

are the axioms for vector norms satisfied?

2.7. Show that for two-dimensional real vectors $\boldsymbol{x}=[x_1, x_2]'$

$$\|\boldsymbol{x}\|=\max\left(|x_1|, |x_2|, \tfrac{2}{3}(|x_1|+|x_2|)\right)$$

satisfies the axioms for vector norms.

Sketch the curve for which $\|\boldsymbol{x}\|=1$.

2.8. Show that all finite-dimensional vector norms are equivalent, i.e., if p_1, p_2 are two such norms then there exist constants p_{12}, p_{21} such that for $\boldsymbol{x}\neq 0$

$$0<p_{12}\leqq\frac{p_1(\boldsymbol{x})}{p_2(\boldsymbol{x})}\leqq p_{21}<\infty.$$

Find best possible values of the p_{ij} when p_i, p_j are any two of the basic norms. In particular show that

$$\|\boldsymbol{x}\|_\infty\leqq\|\boldsymbol{x}\|_1,$$

$$\|\boldsymbol{x}\|_2\leqq\sqrt{n}\,\|\boldsymbol{x}\|_\infty.$$

2.9. Show that the matrix norms (4) and (7) satisfy the four axioms.

2.10. Show that the Frobenius or Schur norm

$$\|\boldsymbol{A}\|_F=\Big[\sum_i\sum_j |a_{ij}|^2\Big]^{1/2}$$

of a matrix $\boldsymbol{A}$ is invariant under orthogonal similarity.

Do any other of the matrix norms discussed have this invariance property?

2.11. If we define

$$v\left(\begin{bmatrix}x_1\\x_2\end{bmatrix}\right)=\begin{cases}\max(|x_1|, |x_2|) & \text{if } x_1\geqq 0\\ \sqrt{x_1^2+x_2^2} & \text{if } x_1<0\end{cases}$$

is v a norm on $\boldsymbol{R}_2$?

Indicate the set of points for which $v(\boldsymbol{x})=1$.

2.12. Why is the adjective "Manhattan" used in (3)?

CHAPTER 3

Induced Norms

In many problems we shall be concerned at the same time with norms of vectors and matrices. It would seem unwise if we used completely unrelated norms for the vectors and matrices. It turns out to be convenient to have a matrix norm "*induced*" by the vector norm. This means that we require a theorem:

Theorem 3.1. *If $n(x)$ is a vector norm satisfying the vector norm axioms then for any matrix* $\mathbf{A}$,

$$m_n(\mathbf{A}) = m(\mathbf{A}) = \sup \frac{n(\mathbf{A}x)}{n(x)},$$

where the supremum is over all non-zero vectors x, satisfies the matrix norm axioms and is called the norm induced by $n(x)$.

It is clear that, no matter what $n(x)$ is, we have

$$m_n(\mathbf{I}) = 1.$$

This implies that (2.7) cannot be an induced norm.

It is not too difficult to determine the matrix norms induced by our three basic vector norms. We discuss them before establishing Theorem 3.1. The results are:

Chebyshev: $\max_i \sum |a_{ij}|$, i.e., max absolute row sum.

Manhattan: $\max_j \sum |a_{ij}|$, i.e., max absolute column sum.

euclidean: [dominant characteristic value of $\mathbf{A}'\mathbf{A}]^{1/2}$.

1. The Chebyshev case

Let $y = \mathbf{A}x$ be the linear transformation given by

$$y_i = \sum a_{ij} x_j;$$

then $\|y\| = \max_i \left|\sum_j a_{ij} x_j\right|$, $\|x\| = \max_i |x_i|$. We have

$$\|y\| \leqq \max_i \sum_j |a_{ij}| \, |x_j| \leqq \max_i \|x\| \sum_j |a_{ij}| = \|x\| \max_i \left(\sum_j |a_{ij}|\right).$$

Hence

(1) $$\|\mathbf{A}\| \leqq \max_i \sum_j |a_{ij}|.$$

We show that we must have equality here. In fact we construct an $\boldsymbol{x}$ such that $\|\boldsymbol{x}\|=1$, $\|\mathbf{A}\boldsymbol{x}\| \geqq \max_i \sum_j |a_{ij}|$ and therefore

(2) $$\|\mathbf{A}\| \geqq \max_i \sum_j |a_{ij}|.$$

Suppose

$$\max_i \sum_j |a_{ij}| = \sum_j |a_{Ij}|,$$

i.e. that the I-th row gives the maximum absolute row sum, and let

$$x_j = \operatorname{sign} a_{Ij}.$$

Then

$$y_I = \sum_j a_{Ij} x_j = \sum |a_{Ij}|.$$

Hence

$$\max |y_i| \geqq |y_I| = \sum |a_{Ij}|.$$

Combining (1) and (2) we have

$$\|\mathbf{A}\| = \max \text{absolute row sum}.$$

This completes the proof.

2. The Manhattan case

The result can be obtained in a similar way. In this case, however, the extremal vector is the unit vector, $\boldsymbol{e}_J$, where

$$\max_j \sum_i |a_{ij}| = \sum_i |a_{iJ}|.$$

3. The euclidean case

The result is somewhat more complicated. We begin with the case when $\mathbf{A}$ is symmetric with characteristic values $\alpha_1, \alpha_2, \ldots, \alpha_n$ where $|\alpha_1| \geqq |\alpha_2| \geqq \ldots \geqq |\alpha_n|$.

We know that we can choose vectors $\boldsymbol{v}_1, \boldsymbol{v}_2, \ldots, \boldsymbol{v}_n$ which are orthogonal and span the whole $\mathbf{R}_n$ and such that $\mathbf{A}\boldsymbol{v}_i = \alpha_i \boldsymbol{v}_i$, $i=1, 2, \ldots, n$. We may also assume these vectors normalized so that $\|\boldsymbol{v}_i\|_2 = 1$, $i=1, 2, \ldots, n$. Any vector $\boldsymbol{x}$ can be expressed as

$$\boldsymbol{x} = c_1 \boldsymbol{v}_1 + c_2 \boldsymbol{v}_2 + \ldots + c_n \boldsymbol{v}_n$$

and we have

$$\mathbf{A}\boldsymbol{x} = \mathbf{A}\left(\sum c_i \boldsymbol{v}_i\right) = \sum c_i \mathbf{A}\boldsymbol{v}_i = \sum c_i \alpha_i \boldsymbol{v}_i.$$

If we assume $\|\boldsymbol{x}\|_2 = 1$ then $\sum c_i^2 = 1$ for $\boldsymbol{x}'\boldsymbol{x} = \left(\sum c_i \boldsymbol{v}_i\right)'\left(\sum c_j \boldsymbol{v}_j\right) = \sum c_i c_j \boldsymbol{v}_i' \boldsymbol{v}_j = \sum c_k^2$.

We note that

$$\|\mathbf{A}\mathbf{x}\|_2^2 = (\mathbf{A}\mathbf{x})'\mathbf{A}\mathbf{x} = \mathbf{x}'\mathbf{A}^2\mathbf{x}$$
$$= (\sum_i c_i \mathbf{x}_i)'(\sum_j c_j \alpha_j^2 \mathbf{x}_j)$$
$$= \sum c_i^2 \alpha_i^2 \quad \text{since} \quad \mathbf{x}_i'\mathbf{x}_j = 0,\ \mathbf{x}_i'\mathbf{x}_i = 1$$

and so

$$\frac{\|\mathbf{A}\mathbf{x}\|_2^2}{\|\mathbf{x}\|_2^2} = \frac{\sum c_i^2 \alpha_i^2}{1} \leqq \alpha_1^2$$

so that

$$\|\mathbf{A}\|_2 \leqq |\alpha_1|.$$

If we take $\mathbf{x} = \mathbf{v}_1$ then

$$\|\mathbf{A}\mathbf{x}\|_2 = |\alpha_1|$$

and so

$$\|\mathbf{A}\|_2 \geqq |\alpha_1|.$$

Hence

$$\|\mathbf{A}\|_2 = |\alpha_1|.$$

In the general case we note that $\mathbf{A}'\mathbf{A}$ is symmetric and we expand in terms of the characteristic vectors $\mathbf{v}_i$ of *this* matrix. If the characteristic values are $p_1, p_2, \ldots, p_n$ where $p_1 \geqq p_2 \geqq \ldots \geqq p_n > 0$ and the $\mathbf{v}_i$ are of unit length we have, when $\mathbf{x} = \sum c_i \mathbf{v}_i$ and $\|\mathbf{x}\|_2 = 1$,

$$\|\mathbf{A}\mathbf{x}\|_2^2 = (\mathbf{A}\mathbf{x})'\mathbf{A}\mathbf{x} = \mathbf{x}'\mathbf{A}'\mathbf{A}\mathbf{x}$$
$$= (\sum c_i \mathbf{v}_i)'(\sum c_j p_j \mathbf{v}_j)$$
$$= \sum c_i^2 p_j$$
$$\leqq p_1.$$

Hence $\|\mathbf{A}\|_2 \leqq \sqrt{p_1}$ and taking $\mathbf{x} = \mathbf{v}_1$ we find $\|\mathbf{A}\|_2 \geqq \sqrt{p_1}$ and so $\|\mathbf{A}\|_2 = \sqrt{p_1}$.

4. Continuity of norms

What do we mean by $f(x)$ continuous at x_0? In the case of one real variable we mean that $f(x)$ is near $f(x_0)$ when x is near x_0. This goes over into the case of a scalar function $f(\mathbf{x})$ of a vector $\mathbf{x}$ provided we know what "$\mathbf{x}$ near $\mathbf{x}_0$" means when these are vectors. We take it to mean that $\|\mathbf{x}-\mathbf{x}_0\|$ is small. But which norm is meant? Actually it does not matter in the finite dimensional case (see Problem 2.8).

But we shall assume that we mean the Chebyshev norm. Thus $f(\mathbf{x})$ is continuous if given any positive ε, there is a δ such that $|f(\mathbf{x})-f(\mathbf{x}_0)| < \varepsilon$ if $\|\mathbf{x}-\mathbf{x}_0\| < \delta$, i.e. if $\mathbf{x}$ lies in a "cube" centered at $\mathbf{x}_0$, side 2δ.

Do we know any continuous functions? Clearly the Chebyshev norm is continuous with $\delta = \varepsilon$. Actually any norm is continuous. This we prove in general. We have, by the triangle axiom

$$|\|\mathbf{x}\| - \|\mathbf{x}_0\|| \leqq \|\mathbf{x}-\mathbf{x}_0\| \leqq \sum \|(\mathbf{x}-\mathbf{x}_0)_i e_i\|.$$

Now by the homogeneity axiom

$$|\|\boldsymbol{x}\| - \|\boldsymbol{x}_0\|| \leqq \sum |(x-x_0)_i| \, \|\boldsymbol{e}_i\| \leqq \|\boldsymbol{x}-\boldsymbol{x}_0\|_\infty \sum \|\boldsymbol{e}_i\|.$$

Now $\sum \|\boldsymbol{e}_i\|$ is a certain constant for each norm. Hence by making $\|\boldsymbol{x}-\boldsymbol{x}_0\|_\infty$ small we can make $|\|\boldsymbol{x}\|-\|\boldsymbol{x}_0\||$ small, which establishes continuity.

In the case of the Manhattan or euclidean norm $\|\boldsymbol{e}_i\| = 1$ and so the constant is n.

We recall the fact that if $f(x)$ is a continuous function of a single real variable then $f(x)$ assumes its bound in any bounded closed interval I, i.e., there is an x_0 such that

$$f(x_0) = \sup_{x \in I} f(x).$$

This result carries over to the case when x is a vector variable.

5. Proof of theorem 3.1

We now discuss the theorem. In the first place, by homogeneity, we can replace

$$\sup_{\boldsymbol{x} \neq 0} \frac{n(\mathbf{A}\boldsymbol{x})}{n(\boldsymbol{x})} \quad \text{by} \quad \sup_{n(\boldsymbol{x})=1} n(\mathbf{A}\boldsymbol{x})$$

since any non-zero $\boldsymbol{x}$ can be written $\boldsymbol{x} = n(\boldsymbol{x})\xi$ where $n(\xi) = 1$.

The set of vectors $\boldsymbol{x}$ for which $n(\boldsymbol{x}) = 1$ is bounded and closed (cf. Problem 3.7). We can therefore replace

$$\sup_{n(\boldsymbol{x})=1} n(\mathbf{A}\boldsymbol{x}) \quad \text{by} \quad \max_{n(\boldsymbol{x})=1} n(\mathbf{A}\boldsymbol{x}).$$

We now show that

$$m(\mathbf{A}) = \max_{n(\boldsymbol{x})=1} n(\mathbf{A}\boldsymbol{x})$$

satisfies the four matrix norm axioms.

(M1) If $\mathbf{A} \neq 0$ we can certainly find an $\boldsymbol{x}$ such that $n(\boldsymbol{x}) = 1$ and $\mathbf{A}\boldsymbol{x} \neq 0$ and so $n(\mathbf{A}\boldsymbol{x}) \neq 0$ and so $m(\mathbf{A}) \neq 0$. Clearly $m(\mathbf{A}) \geqq 0$. If $\mathbf{A} = 0$ then $\mathbf{A}\boldsymbol{x} = 0$ and so $n(\mathbf{A}\boldsymbol{x}) = 0$ and $m(\mathbf{A}) = 0$.

(M2) $m(a\mathbf{A}) = \max n(a\mathbf{A}\boldsymbol{x}) = \max |a| n(\mathbf{A}\boldsymbol{x}) = |a| \max n(\mathbf{A}\boldsymbol{x}) = |a| \, \|\mathbf{A}\|$.

(M3) Consider $m(\mathbf{A}+\mathbf{B})$. There is an $\boldsymbol{x}$ such that $n(\boldsymbol{x}) = 1$ and

$$m(\mathbf{A}+\mathbf{B}) = n((\mathbf{A}+\mathbf{B})\boldsymbol{x}).$$

Now

$$\begin{aligned} m(\mathbf{A}+\mathbf{B}) &= n((\mathbf{A}+\mathbf{B})\boldsymbol{x}) = n(\mathbf{A}\boldsymbol{x}+\mathbf{B}\boldsymbol{x}) \\ &\leqq n(\mathbf{A}\boldsymbol{x}) + n(\mathbf{B}\boldsymbol{x}) \quad \text{(triangle axiom)} \\ &\leqq m(\mathbf{A}) + m(\mathbf{B}) \quad \text{(definition of } m\text{)}. \end{aligned}$$

(M4) Consider $m(\boldsymbol{AB})$. There is an $\boldsymbol{x}$ such that $n(\boldsymbol{x})=1$ and

$$m(\boldsymbol{AB})=n(\boldsymbol{ABx})$$

and so

$$m(\boldsymbol{AB})=n(\boldsymbol{ABx})\leqq m(\boldsymbol{A})n(\boldsymbol{Bx})\leqq m(\boldsymbol{A})m(\boldsymbol{B})n(\boldsymbol{x})=m(\boldsymbol{A})m(\boldsymbol{B}).$$

This completes the proof of Theorem 3.1.

One of the uses of this theorem is the fact that it gives

$$n(\boldsymbol{Ax})\leqq m(\boldsymbol{A})n(\boldsymbol{x})$$

or, as it usually is written,

$$\|\boldsymbol{Ax}\|_n\leqq\|\boldsymbol{A}\|_m\,\|\boldsymbol{x}\|_n. \tag{3}$$

This inequality is a best possible one: our proof shows that we can always find a vector $\boldsymbol{x}$ for which equality is attained.

In many applications, since it is only (3) that we use, we can take any matrix norm $\|\cdot\|$ for which

$$\|\boldsymbol{Ax}\|_n\leqq\|\boldsymbol{A}\|\,\|\boldsymbol{x}\|_n$$

is true for all $\boldsymbol{x}$ with the given vector norm $\|\cdot\|_n$. Such a matrix norm is said to be *compatible with* or *subordinate to* the vector norm.

6. Spectral radius and convergence

Another important quantity associated with a matrix $\boldsymbol{A}$ is its *spectral radius* $\varrho(\boldsymbol{A})$: if $\alpha_1, \ldots, \alpha_n$ are the characteristic values of $\boldsymbol{A}$ then

$$\varrho(\boldsymbol{A})=\max|\alpha_i|.$$

The spectral radius of a matrix is not a norm. For instance, if

$$\boldsymbol{A}=\begin{bmatrix}0&1\\0&0\end{bmatrix},\quad \boldsymbol{B}=\begin{bmatrix}0&0\\1&0\end{bmatrix}$$

we have $\varrho(\boldsymbol{A})=0$, $\varrho(\boldsymbol{B})=0$ but $\varrho(\boldsymbol{A}+\boldsymbol{B})=1$ so that the triangle axiom is not satisfied.

Once we have "norms" available we can discuss convergence. We shall establish

Theorem 3.2. *If* $\|\boldsymbol{A}\|=\theta<1$ *then*

$$(\boldsymbol{I}-\boldsymbol{A})^{-1}=\boldsymbol{I}+\boldsymbol{A}+\boldsymbol{A}^2+\ldots$$

where convergence of the series on the right is in the sense of the matrix norm.

Proof. The relation $(\boldsymbol{I}-\boldsymbol{A})^{-1}=\boldsymbol{I}+\boldsymbol{A}+\boldsymbol{A}^2+\ldots$ means that $\boldsymbol{I}-\boldsymbol{A}$ is non-singular and that

$$\|(\boldsymbol{I}-\boldsymbol{A})^{-1}-(\boldsymbol{I}+\boldsymbol{A}+\ldots+\boldsymbol{A}^{r-1})\|\to 0 \quad \text{as} \quad r\to\infty.$$

Write $\boldsymbol{S}_r=\boldsymbol{I}+\boldsymbol{A}+\ldots+\boldsymbol{A}^{r-1}$. Then, as in the scalar case,

$$\boldsymbol{A}\boldsymbol{S}_r=\boldsymbol{A}+\boldsymbol{A}^2+\ldots+\boldsymbol{A}^{r-1}+\boldsymbol{A}^r \quad \text{so that}$$

$$(\boldsymbol{I}-\boldsymbol{A})\boldsymbol{S}_r=\boldsymbol{I}-\boldsymbol{A}^r. \tag{4}$$

If $\|\boldsymbol{A}\|<1$ then $\boldsymbol{I}-\boldsymbol{A}$ is non-singular for if it were, then 1 would be a characteristic value of $\boldsymbol{A}$ and hence the spectral radius $\varrho(\boldsymbol{A})\geqq 1$ but (Problem 3.1) $\|\boldsymbol{A}\|\geqq\varrho(\boldsymbol{A})$ which gives a contradiction. We can therefore rewrite (4) as

$$\boldsymbol{S}_r-(\boldsymbol{I}-\boldsymbol{A})^{-1}=-(\boldsymbol{I}-\boldsymbol{A})^{-1}\boldsymbol{A}^r$$

and this gives, by submultiplicativity,

$$\|\boldsymbol{S}_r-(\boldsymbol{I}-\boldsymbol{A})^{-1}\|\leqq\|(\boldsymbol{I}-\boldsymbol{A})^{-1}\|\,\|\boldsymbol{A}\|^r\leqq\|(\boldsymbol{I}-\boldsymbol{A})^{-1}\|\theta^r\to 0.$$

Hence, if $\|\boldsymbol{A}\|=\theta<1$ for *any* particular norm

$$(\boldsymbol{I}-\boldsymbol{A})^{-1}=\boldsymbol{I}+\boldsymbol{A}+\boldsymbol{A}^2+\ldots$$

convergence being in the sense of that norm.

We now relate Theorem 3.2 to

Theorem 3.3. *If* $\varrho(\boldsymbol{A})<1$ *then*

$$(\boldsymbol{I}-\boldsymbol{A})^{-1}=\boldsymbol{I}+\boldsymbol{A}+\boldsymbol{A}^2+\ldots$$

where convergence is in the sense of a certain norm.

Proof. The simplest way to establish this is to use the fact that there are matrix norms such that $\|\boldsymbol{A}\|<\varrho(\boldsymbol{A})+\varepsilon$ for any $\varepsilon>0$, and then to appeal to Theorem 3.2.

However to establish the fact used, we have to employ the ε-version of the Jordan normal form. We discuss this first. It is well-known that there are matrices which are not similar to a diagonal matrix. For instance

$$\begin{bmatrix}0 & 1\\ 0 & 0\end{bmatrix}=\begin{bmatrix}\delta & -\beta\\ \gamma & \alpha\end{bmatrix}\begin{bmatrix}0 & 0\\ 0 & 0\end{bmatrix}\begin{bmatrix}\alpha & \beta\\ \gamma & \delta\end{bmatrix},\quad \alpha\delta-\beta\gamma=1,$$

is clearly impossible. However

$$\begin{bmatrix}0 & 1\\ 0 & 0\end{bmatrix}=\begin{bmatrix}\delta & -\beta\\ -\gamma & \alpha\end{bmatrix}\begin{bmatrix}0 & \varepsilon\\ 0 & 0\end{bmatrix}\begin{bmatrix}\alpha & \beta\\ \gamma & \delta\end{bmatrix},\quad \alpha\delta-\beta\gamma=1$$

is possible for any $\varepsilon\neq 0$-indeed we take $\alpha=\delta^{-1}=\varepsilon^{1/2}$, $\gamma=0$, β arbitrary. The usual version of the Jordan theorem states that any matrix $\boldsymbol{A}$ is similar to a matrix having for its diagonal the characteristic values of $\boldsymbol{A}$ and having all other elements zero, except perhaps those immediately adjacent to the diagonal on its right, which are 1. An additional similarity will permit these 1's to be changed to any non-zero numbers, in particular to arbitrarily small $\varepsilon \neq 0$. (This has just been exemplified in the 2×2 case.)

This result being granted we proceed as follows. Let $\mathbf{S}$ be the matrix which produces the ε-Jordan normal form for $\mathbf{A}$ say, $\mathbf{B}=\mathbf{S}^{-1}\mathbf{A}\mathbf{S}$. Write

$$\|\mathbf{x}\| = \|\mathbf{S}^{-1}\mathbf{x}\|_\infty;$$

it is easy to verify that $\|\cdot\|$ is a vector norm. We compute the matrix norm it induces:

$$\begin{aligned}\|\mathbf{A}\| &= \max_{\mathbf{x}\neq\mathbf{0}} [\|\mathbf{A}\mathbf{x}\|/\|\mathbf{x}\|] \\ &= \max_{\mathbf{x}\neq\mathbf{0}} [\|\mathbf{S}^{-1}\mathbf{A}\mathbf{S}\mathbf{S}^{-1}\mathbf{x}\|_\infty/\|\mathbf{S}^{-1}\mathbf{x}\|_\infty] \\ &= \max_{\mathbf{y}\neq\mathbf{0}} [\|\mathbf{B}\mathbf{y}\|_\infty/\|\mathbf{y}\|_\infty] \\ &= \|\mathbf{B}\|_\infty .\end{aligned}$$

(Here we have written $\mathbf{y}=\mathbf{S}^{-1}\mathbf{x}$ and used the fact that $\mathbf{S}$ is non-singular so that the set of vectors $\mathbf{x}\neq\mathbf{0}$ is identical with the set $\mathbf{y}\neq\mathbf{0}$.)

It is easy to estimate $\|\mathbf{B}\|_\infty$. Generally, $\|\mathbf{M}\|_\infty$ is the maximum absolute row sum of $\mathbf{M}$. Since the only non-zero elements in the rows of $\mathbf{B}$ are α_i and at most one ε we have

$$\begin{aligned}\|\mathbf{B}\|_\infty &\leqq \max |\alpha_i| + \varepsilon \\ &= \varrho(\mathbf{A}) + \varepsilon.\end{aligned}$$

This is the result we need.

7. The matrix norm induced by the p-norm

It is natural to ask whether we can give a formula for the matrix norm induced by the vector norm $\|\cdot\|_p$. Such a result does not appear to be available but we can give an upper bound for the matrix norm which is exact in the extreme cases $p=1$, $p=\infty$. In practice such an upper bound is usually adequate for the purposes of error estimation.

We use some elementary properties of convex functions. The original definition is that $f(t)$ is convex, in [0, 1], say, if for any t_1, t_2 in [0, 1] we have

$$f(\tfrac{1}{2}t_1+\tfrac{1}{2}t_2) \leqq \tfrac{1}{2}f(t_1)+\tfrac{1}{2}f(t_2),$$

i.e., the curve is below the chord. It can be shown that this result can be extended to the following: if $t_1, t_2, \ldots, t_n$ are in [0, 1] and if $p_1, p_2, \ldots, p_n$ are positive weights with $\sum p_i = 1$ then

$$f(\sum p_i t_i) \leqq \sum p_i f(t_i).$$

We also note that $f(t)=t^p$ is convex in [0, 1] for $p\geqq 1$.

Consider the transformation $\boldsymbol{y}=\boldsymbol{A}\boldsymbol{x}$ and denote by $r_1, r_2, \ldots, r_n$ and $c_1, c_2, \ldots, c_n$ the row and column sums of $|\boldsymbol{A}|$. Observe that

$$\frac{|y_i|}{r_i} \leqq \sum_{j=1}^{n} \frac{|a_{ij}|}{r_i} |x_j|$$

and that the right hand side is a weighted mean of $|x_1|, \ldots, |x_n|$. It follows, from the remarks in the preceding paragraph, that

$$\left[\frac{|y_i|}{r_i}\right]^p \leqq \left[\sum_{j=1}^{n} \frac{|a_{ij}|}{r_i} |x_j|\right]^n \leqq \sum_{j=1}^{n} \frac{|a_{ij}|}{r_i} |x_j|^p,$$

so that

$$|y_i|^p \leqq r_i^{p-1} \sum_{j=1}^{n} |a_{ij}|\,|x_j|^p \leqq [\max_i r_i]^{p-1} \sum_{j=1}^{n} |a_{ij}|\,|x_j|^p.$$

Summing with respect to i we get

$$\|\boldsymbol{y}\|_p^p \leqq [\max_i r_i]^{p-1} \sum_{i=1}^{n} \sum_{j=1}^{n} |a_{ij}|\,|x_j|^p \leqq [\max_i r_i]^{p-1} [\max_j c_j]\, \|\boldsymbol{x}\|_p^p$$

and, taking p-th roots,

$$\|\boldsymbol{y}\|_p \leqq [\max_i r_i]^{1-1/p} [\max_j c_j]^{1/p}\, \|\boldsymbol{x}\|_p$$

which gives

$$\|\boldsymbol{A}\|_{p,\,p} = \sup_{x \neq 0} \frac{\|\boldsymbol{y}\|_p}{\|\boldsymbol{x}\|_p} \leqq [\max_i r_i]^{1-1/p} [\max_j c_j]^{1/p}.$$

This is the bound given by Tosio Kato. For $p=\infty$ the second factor on the right drops out and for $p=1$ the first factor drops. As a byproduct in the case $p=2$ we find

$$\begin{aligned} \|\boldsymbol{A}\| &= (\text{dominant characteristic value of } \boldsymbol{A}\boldsymbol{A}')^{1/2} \\ &\leqq (\max_i r_i \max_j c_j)^{1/2}. \end{aligned}$$

Chapter 3, Problems

3.1. Applying the inequality

$$\|\boldsymbol{A}\boldsymbol{x}\| \leqq \|\boldsymbol{A}\|\,\|\boldsymbol{x}\|$$

for compatible norms and any matrix $\boldsymbol{A}$, when $\boldsymbol{x}$ is a characteristic vector of $\boldsymbol{A}$, show that

$$\varrho(\boldsymbol{A}) \leqq \|\boldsymbol{A}\|.$$

3.2. What are the norms of the matrix $\boldsymbol{A}=\begin{bmatrix}1 & 2\\ 3 & 4\end{bmatrix}$ which are induced by the euclidean, the Chebyshev and the Manhattan norms?

In each of the three cases find all vectors $\boldsymbol{x}$ satisfying

$$\|\boldsymbol{A}\boldsymbol{x}\| = \|\boldsymbol{A}\|\,\|\boldsymbol{x}\|.$$

What is the spectral radius of $\boldsymbol{A}$?

3.3. In the definition of induced norm we have used the same norm for the vector $\boldsymbol{y}=\boldsymbol{A}\boldsymbol{x}$ as for the vector $\boldsymbol{x}$. This is not necessary, and it is easy to show that if $n_1(\boldsymbol{x})$ and $n_2(\boldsymbol{x})$ are two vector norms then

$$\|\boldsymbol{A}\|_{12} = \sup_{\boldsymbol{x}\neq 0} \frac{n_1(\boldsymbol{A}\boldsymbol{x})}{n_2(\boldsymbol{x})}$$

is a not necessarily submultiplicative matrix norm. Calculate

$$\|\boldsymbol{A}\|_{12}$$

in terms of the elements of $\boldsymbol{A}$ when

$n_1(\boldsymbol{x})$ is the Chebyshev norm

and

$n_2(\boldsymbol{x})$ is the Manhattan norm.

3.4. Draw the curves for which $\|\boldsymbol{x}\|_1=1$ and for which $\|\boldsymbol{x}\|_\infty=1$, where $\boldsymbol{x}=[x_1, x_2]'$ is a real two-dimensional vector.

Consider the norm $s(\boldsymbol{x})$ on vectors $\boldsymbol{x}$ for which $s(\boldsymbol{x})=1$ holds on the polygon formed by joining the points

$$(-1, 0),\ (0, 1),\ (1, 1),\ (1, 0),\ (0, -1),\ (-1, -1).$$

Find constants c_1, c_2, d_1, d_2 such that the inequalities

$$c_1\|\boldsymbol{x}\|_\infty \leqq s(\boldsymbol{x}) \leqq c_2\|\boldsymbol{x}\|_\infty$$

$$d_1\|\boldsymbol{x}\|_1 \leqq s(\boldsymbol{x}) \leqq d_2\|\boldsymbol{x}\|_1$$

hold sharply, giving in each case a vector where equality holds.

If $S(\boldsymbol{A})$ is the norm induced on 2×2 real matrices by $s(\boldsymbol{x})$ find

$$S\left(\begin{bmatrix} 1 & 2 \\ 3 & 4 \end{bmatrix}\right)$$

and find a vector $\boldsymbol{x}$ such that $s(\boldsymbol{x})=1$ and

$$S([x_1+2x_2,\ 3x_1+4x_2]) = S\left(\begin{bmatrix} 1 & 2 \\ 3 & 4 \end{bmatrix}\right).$$

3.5. Show that the two matrix norms $\|\boldsymbol{A}\|_F = \sqrt{\sum_{i,j} |a_{ij}|^2}$ and $\|\boldsymbol{A}\|_M = n \max_{i,j} |a_{ij}|$

are consistent or compatible with the euclidean vector norm $\|\boldsymbol{x}\|_2$, i.e., that

$$\|\mathbf{A}\boldsymbol{x}\|_2 \leqq m(\mathbf{A})\,\|\boldsymbol{x}\|_2,$$

where m is either of the two norms.

3.6. Show that if m is any submultiplicative matrix norm and $\boldsymbol{\eta}\neq 0$ is any vector then $n(\boldsymbol{x})=m(\boldsymbol{x}\boldsymbol{\eta}')$ is a vector norm with which m is compatible.

3.7. Justify in detail, in the real two-dimensional case, the following statement on p. 22:

"The set of vectors for which $n(\boldsymbol{x})=1$ is bounded and closed."

3.8. Find $\|\mathbf{A}\|_\infty$, $\|\mathbf{A}\|_2$ in the case when

$$\mathbf{A}=\mathbf{A}_8=\begin{bmatrix}-2 & 1 & 0 & 0\\ 1 & -2 & 1 & 0\\ 0 & 1 & -2 & 1\\ 0 & 0 & 1 & -2\end{bmatrix}$$

and find $\|\mathbf{A}\|_1$ in the case when

$$\mathbf{A}=\mathbf{A}_{16}=\begin{bmatrix}1 & 0 & 0 & 0\\ 1 & 2 & 0 & 0\\ 1 & 3 & 3 & 0\\ 1 & 4 & 6 & 4\end{bmatrix}$$

In each case find a vector $\boldsymbol{x}\neq 0$ for which $\|\mathbf{A}\boldsymbol{x}\|=\|\mathbf{A}\|\,\|\boldsymbol{x}\|$.

CHAPTER 4

The Inversion Problem I: Theoretical Arithmetic

The writing of *efficient* programs for the more complicated operations on matrices is a truly professional job, still going on, despite the investment of many millions of dollars. We shall not attempt to compete, except in very simple cases, but we shall use *critically* the software prepared for us by our colleagues. We define efficiency roughly as fast and accurate; with comparatively small scale problems on which we must work, our examination of speed must be largely theoretical. But we can examine practically the question of accuracy.

We begin with three basic but related problems:

(1) evaluate $\det \mathbf{A}$,
(2) solve $\mathbf{A}\mathbf{x}=\mathbf{b}$,
(3) evaluate $\mathbf{A}^{-1}$,

and in this chapter we shall be discussing, on the whole, the speed of solution processes. We shall assume that all our arithmetic operations are done *without error* — this is what we mean by *theoretical arithmetic* in contrast with *practical computation* where we take into account the round-off errors which usually occur.

In this context we can always find $\det \mathbf{A}$ and we can solve the system $\mathbf{A}\mathbf{x}=\mathbf{b}$ or invert $\mathbf{A}$ if $\mathbf{A}$ is non-singular. These facts are established in any treatment of linear algebra — we have to look at the treatments from a constructive and practical point of view.

We begin by discussing special cases.

1. Diagonal Matrix

If $\mathbf{A}=\mathbf{D}=\operatorname{diag}[a_{11}, \ldots, a_{nn}]$ is non-singular then the solution to $\mathbf{A}\mathbf{x}=\mathbf{b}$ is obtained directly: $x_i=b_i/a_{ii}$; also

$$\det \mathbf{A}=\prod a_{ii},$$
$$\mathbf{A}^{-1}=\operatorname{diag}[a_{11}^{-1}, \ldots, a_{nn}^{-1}].$$

The solutions are obtained at the expense of n divisions, the determinant at the expense of $n-1$ multiplications and the inverse by n divisions.

2. Triangular Matrix

Suppose $\mathbf{A}$ has $a_{ij}=0$ if $i>j$. Then $\det \mathbf{A}=\prod a_{ii}$, and we can obtain the x_i's by "back substitution":

$$x_n = b_n/a_{nn}$$

$$x_{n-1} = [b_{n-1} - a_{n-1,n}x_n]/a_{n-1,n-1}$$

$$\dots$$

$$x_1 = \left[b_1 - \sum_{j=2}^{n} a_{ij}x_j\right]/a_{11}.$$

The expense of this in multiplications (and divisions) and additions (and subtractions) can be easily estimated. To find x_1 involves $n-1$ multiplications, one division and $n-1$ additions. In all we will have about $n^2/2$ multiplications and about the same number of additions. The evaluation of the determinant still requires only $n-1$ multiplications. The inversion problem is given as Problem 4.16.

3. Triple Diagonal Matrix

We discuss next the case of a triple diagonal system — this is a basic problem in numerical linear algebra. It arises, e.g., in approximate solution of second order differential equations, and it has proved convenient to reduce certain characteristic value problems to this case.

We change our notation for convenience:

$$\begin{aligned}
a_1x_1 + b_1x_2 \qquad\qquad\qquad\qquad &= d_1\\
c_2x_1 + a_2x_2 + b_2x_3 \qquad\qquad &= d_2\\
\dots\qquad\qquad\qquad\qquad&\\
c_{n-1}x_{n-2} + a_{n-1}x_{n-1} + b_{n-1}x_n &= d_{n-1}\\
c_nx_{n-1} + a_nx_n &= d_n.
\end{aligned}$$

The determinant, φ_n, of the $n\times n$ triple diagonal system is easily evaluated by a three-term recurrence. If we write $\varphi_0=1$, we find $\varphi_1=a_1$, $\varphi_2=a_1a_2-b_1c_2=$ $=a_2\varphi_1-b_1c_2\varphi_0$ and generally, expanding by minors of the last row,

$$\varphi_r = a_r\varphi_{r-1} - b_{r-1}c_r\varphi_{r-2}.$$

Thus φ_n can be computed at the expense of about $3n$ multiplications and n additions.

We now discuss the linear equation problem. It is clear that we can find x_1 in terms of x_2 from the first equation. If we substitute this value of x_1 in the second we can find x_3 in terms of x_2. Proceeding in this way, using the first $(n-1)$ equations, we get relations of the form

$$(4)_r \qquad x_r = f_r x_{r+1} + g_r$$

for $r=1, 2, \ldots, n-1$. [Here f_r, g_r are yet to be determined.] We can now substitute in the last equation and solve for x_n. We then proceed by back substitution in (4) to get $x_{n-1}, x_{n-2}, \ldots, x_1$ successively.

Let us describe this process more precisely. There are a few degenerate cases which ought to be dealt with separately but which we will disregard. For instance, if $a_1=0$ we obtain x_2 immediately and the system is reduced; again, if the system is symmetrical and if any b is zero the system splits up into two "uncoupled" systems.

Assume $(4)_{r-1}$ and substitute in the r-th equation

$$c_r x_{r-1} + a_r x_r + b_r x_{r+1} = d_r$$

to get

$$(5)_r \qquad x_r = \frac{-b_r}{c_r f_{r-1} + a_r} x_{r+1} + \frac{d_r - c_r g_{r-1}}{c_r f_{r-1} + a_r}.$$

Identifying this with $(4)_r$ we see that f_r, g_r satisfy the recurrence relations:

$$f_r = \frac{-b_r}{c_r f_{r-1} + a_r}, \quad g_r = \frac{d_r - c_r g_{r-1}}{c_r f_{r-1} + a_r}, \qquad r = 1, 2, \ldots, n-1.$$

Observe that $c_1=0$, $b_n=0$ and that we need not define f_0, g_0 in the first equation $(5)_1$ which is, correctly,

$$x_1 = \frac{-b_1}{a_1} x_2 + \frac{d_1}{a_1}.$$

From these we compute $f_1, \ldots, f_n$ and $g_1, \ldots, g_n$ at the cost of about $3n$ additions, $2n$ multiplications and $2n$ divisions. The last equation $(5)_n$ gives $f_n=0$ and so $x_n=g_n$ and then we obtain $x_{n-1}, \ldots, x_1$ by back substitution in (4) at the cost of n multiplications and n additions.

Observe that the inverse of a triple diagonal matrix can be a full matrix. (Cf. Problem 5.13(iv).)

4. Band Matrices

We have just seen that problems involving triple diagonal matrices can be handled cheaply. The ideas just exposed can be adapted to the case of 5-diagonal matrices, which turn up in the solution of fourth order differential equations, and more generally, to band matrices where

$$a_{i,j}=0 \quad \text{if} \quad |i-j| \geqq m, \quad m \ll n.$$

Specially designed programs for handling such cases are usually available in any computing system and should be used in preference to general programs.

We note that in many cases, including those arising from differential equations, iterative methods [e.g. those of Jacobi, Gauss—Seidel, Young (Successive Over-Relaxation), and the Alternating Direction Implicit Methods] may be more efficient. Some of these are discussed in Chapter 9.

5. The general case

We now want to discuss one of the many variants of the Gaussian elimination method. We do this in matrix notation and shall make use of "partitioned" or "block" matrices. It can be proved that we can operate on block matrices, provided the partitioning is appropriate, just as an ordinary matrices, but paying attention of course to the order of factors. The interpretation of block matrices as linear transformations is clear: the blocks act on subspaces.

The formal justification of the operations on block matrices depends on the fact that finite sums can be rearranged and is rather uncivilized. (Cf. Faddeev and Sominskii.) We outline the argument in a special case.

$$\boldsymbol{A}=\begin{bmatrix} \boldsymbol{A}_{11} & \boldsymbol{A}_{12} \\ \boldsymbol{A}_{21} & \boldsymbol{A}_{22} \end{bmatrix}, \quad \boldsymbol{B}=\begin{bmatrix} \boldsymbol{B}_{11} & \boldsymbol{B}_{12} \\ \boldsymbol{B}_{21} & \boldsymbol{B}_{22} \end{bmatrix}, \quad \boldsymbol{C}_{+}=\boldsymbol{A}+\boldsymbol{B}, \quad \boldsymbol{C}_{\times}=\boldsymbol{AB}.$$

We have to show that

$$\boldsymbol{C}_{+}=\begin{bmatrix} \boldsymbol{A}_{11}+\boldsymbol{B}_{11}, & \boldsymbol{A}_{12}+\boldsymbol{B}_{12} \\ \boldsymbol{A}_{21}+\boldsymbol{B}_{21}, & \boldsymbol{A}_{22}+\boldsymbol{B}_{22} \end{bmatrix}, \quad \boldsymbol{C}_{\times}=\begin{bmatrix} \boldsymbol{A}_{11}\boldsymbol{B}_{11}+\boldsymbol{A}_{12}\boldsymbol{B}_{21}, & \boldsymbol{A}_{11}\boldsymbol{B}_{12}+\boldsymbol{A}_{12}\boldsymbol{B}_{22} \\ \boldsymbol{A}_{21}\boldsymbol{B}_{11}+\boldsymbol{A}_{22}\boldsymbol{B}_{21}, & \boldsymbol{A}_{21}\boldsymbol{B}_{12}+\boldsymbol{A}_{22}\boldsymbol{B}_{22} \end{bmatrix}.$$

The addition result is trivial. To deal with the multiplication result, observe that a typical element in $\boldsymbol{AB}$ is an inner product:

$$[-----**********]\begin{bmatrix} + \\ + \\ + \\ + \\ + \\ \circ \\ \circ \\ \circ \\ \circ \\ \circ \\ \circ \\ \circ \\ \circ \\ \circ \\ \circ \end{bmatrix}$$

and this can be represented as the sum of the two inner products

$$[-----]\begin{bmatrix} + \\ + \\ + \\ + \\ + \end{bmatrix} \quad \text{and} \quad [**********]\begin{bmatrix} \circ \\ \circ \\ \circ \\ \circ \\ \circ \\ \circ \\ \circ \\ \circ \\ \circ \\ \circ \end{bmatrix},$$

which is what appears in $\boldsymbol{C}_{\times}$.

Lemma. *If* $\mathbf{C}=\mathbf{AB}$ *is non-singular, so are* $\mathbf{A}$ *and* $\mathbf{B}$.

Theorem 4. 1. *If all the leading submatrices of an* $n\times n$ *matrix* $\mathbf{A}$ *are non-singular, then there is a representation of* $\mathbf{A}$ *as a product*

$$\mathbf{A}=\mathbf{LU}$$

where $\mathbf{L}$ *is a lower triangular matrix and* $\mathbf{U}$ *an upper triangular one.*

Proof. We use induction. For $n=1$ the result is trivial: we can write

$$[a_{11}]=[b_{11}][c_{11}]$$

and one of the b_{11} or c_{11} can be chosen arbitrarily (different from zero).

Assume the result for $(n-1)\times(n-1)$ matrices. Partition the $n\times n$ matrix $\mathbf{A}$ as

$$\mathbf{A}=\begin{bmatrix}\mathbf{A}_{n-1} & \boldsymbol{u}\\ \boldsymbol{v}' & a_{nn}\end{bmatrix}$$

where $\boldsymbol{u}, \boldsymbol{v}$ are $(n-1)$-dimensional vectors.

We want to get

$$\mathbf{A}=\mathbf{LU} \quad \text{where} \quad \mathbf{L}=\begin{bmatrix}\mathbf{L}_{n-1} & 0\\ \boldsymbol{x}' & l_{nn}\end{bmatrix}, \quad \mathbf{U}=\begin{bmatrix}\mathbf{U}_{n-1} & \boldsymbol{y}\\ 0 & u_{nn}\end{bmatrix}$$

where $\mathbf{L}_{n-1}$, $\mathbf{U}_{n-1}$, $\boldsymbol{x}$, $\boldsymbol{y}$, l_{nn}, u_{nn} are to be determined. Multiplying out these block matrices we see that we want to have

$$\begin{bmatrix}\mathbf{A}_{n-1} & \boldsymbol{u}\\ \boldsymbol{v}' & a_{nn}\end{bmatrix}=\begin{bmatrix}\mathbf{L}_{n-1}\mathbf{U}_{n-1}, & \mathbf{L}_{n-1}\boldsymbol{y}\\ \boldsymbol{x}'\mathbf{U}_{n-1}, & \boldsymbol{x}'\boldsymbol{y}+l_{nn}u_{nn}\end{bmatrix}.$$

In virtue of our induction assumption, $\mathbf{L}_{n-1}$, $\mathbf{U}_{n-1}$ exist (because all the leading submatrices of $\mathbf{A}_{n-1}$, being included among those of $\mathbf{A}$, are non-singular). Our lemma shows that $\mathbf{L}_{n-1}$, $\mathbf{U}_{n-1}$ are non-singular and so

$$\boldsymbol{u}=\mathbf{L}_{n-1}\boldsymbol{y} \quad \text{and} \quad \boldsymbol{v}=\boldsymbol{x}\mathbf{U}_{n-1}$$

can be solved to get

$$\boldsymbol{x}=\boldsymbol{v}\mathbf{U}_{n-1}^{-1}, \quad \boldsymbol{y}=\mathbf{L}_{n-1}^{-1}\boldsymbol{u}.$$

We obtain, finally,

$$l_{nn}u_{nn}=a_{nn}-\boldsymbol{x}'\boldsymbol{y}$$

and assigning one of these arbitrarily the other is fixed.

Various conventions are possible to remove the arbitrariness. We could, e.g., assume that $\mathbf{L}$ or $\mathbf{U}$ had a unit diagonal or we could assume that $l_{nn}=u_{nn}$ (this relates our decomposition to the "squareroot method") or we could assume both $\boldsymbol{L}$ and $\boldsymbol{U}$ had unit diagonals in which case our decomposition would be of the form:

$$\boldsymbol{A}=\boldsymbol{L}\boldsymbol{D}\boldsymbol{U}.$$

A. M. Turing used the last normalization and called the theorem the "$\boldsymbol{LDU}$ theorem".

6. Comments

(1) The assumption that *"the leading submatrices are non-singular"* is necessary. Suppose we had

$$\boldsymbol{A}=\begin{bmatrix}0 & 1\\ 1 & 0\end{bmatrix}=\begin{bmatrix}a & 0\\ b & c\end{bmatrix}\begin{bmatrix}d & e\\ 0 & f\end{bmatrix}.$$

Then $ad=0$, $ae=0$, $bd=1$, $be+cf=0$. The first condition gives $a=0$ or $d=0$; if $a=0$ we cannot satisfy the second condition while if $d=0$ we cannot satisfy the third condition. Thus a triangular factorization of $\boldsymbol{A}$ is impossible. However, if we interchange the first and second rows of $\boldsymbol{A}$ it becomes $\boldsymbol{I}$ and factorization is trivial. This remark can be elaborated to deal with the general case. Indeed we can show that *if $\boldsymbol{A}$ is non-singular then there is a permutation matrix $\boldsymbol{P}$ such that $\boldsymbol{PA}$ satisfies the condition quoted.*

We prove this by induction. The result is trivial in the case $n=1$. For general n expand $\det \boldsymbol{A}$ according to minors of its last column:

$$0\neq\det\boldsymbol{A}=a_{nn}\det\boldsymbol{A}_{nn}-a_{n-1,n}\det\boldsymbol{A}_{n-1,n}+\ldots+(-1)^{n}a_{1n}\det\boldsymbol{A}_{1n}.$$

At least one of the matrices $\boldsymbol{A}_{in}$ is therefore non-singular. Let $\boldsymbol{P}_1$ be the permutation which transfers $\boldsymbol{A}_{in}$ to the top left corner and replaces the n-th row by the i-th row. By the induction hypothesis there is a $n-1\times n-1$ permutation matrix $\hat{\boldsymbol{P}}$ such that the matrix

$$\begin{bmatrix}\hat{\boldsymbol{P}} & 0\\ 0 & 1\end{bmatrix}\boldsymbol{P}_1\boldsymbol{A}$$

has its leading submatrices of order $1, 2, \ldots, n-1$ all non-singular, and of course this matrix itself is necessarily non-singular having the same determinant as $\boldsymbol{A}$. The product

$$\begin{bmatrix}\hat{\boldsymbol{P}} & 0\\ 0 & 1\end{bmatrix}\boldsymbol{P}_1$$

is the permutation matrix sought.

(2) If $\boldsymbol{A}$ is symmetric we can obtain a representation of the form $\boldsymbol{A}=\boldsymbol{L}\boldsymbol{L}'$ or $\boldsymbol{A}=\boldsymbol{L}\boldsymbol{D}\boldsymbol{L}'$. To prove this either repeat the induction proof as in the general case, or transpose and use uniqueness. (Cf. p. 59.)

This representation is often called the Cholesky factorization.

(3) This theorem implies the Gaussian reduction: we get from $\boldsymbol{A}\boldsymbol{x}=\boldsymbol{b}$ to $\boldsymbol{U}\boldsymbol{x}=\boldsymbol{c}$ by premultiplication of $\boldsymbol{A}\boldsymbol{x}=\boldsymbol{b}$ by $\boldsymbol{L}^{-1}$ to get $\boldsymbol{L}^{-1}\boldsymbol{L}\boldsymbol{U}\boldsymbol{x}=\boldsymbol{L}^{-1}\boldsymbol{b}$, and we can assume $\boldsymbol{L}$ to have units on the diagonal.

(4) This theorem implies the Gram—Schmidt result. Let $\boldsymbol{F}$ be a set of n linearly independent vectors $\boldsymbol{f}_1, \boldsymbol{f}_2, \dots, \boldsymbol{f}_n$ in n dimensions. We want a linearly independent set $\boldsymbol{\Phi}$ of orthonormal vectors $\varphi_1, \varphi_2, \dots, \varphi_n$ which are a linear transformation of the $\boldsymbol{f}$'s.

We can write

$$\boldsymbol{F}=\begin{bmatrix} f_1^{(1)} & & f_n^{(1)} \\ f_1^{(2)} & \dots & f_n^{(2)} \\ \vdots & & \vdots \\ f_1^{(n)} & & f_n^{(n)} \end{bmatrix}, \quad \boldsymbol{\Phi}=\begin{bmatrix} \varphi_1^{(1)} & & \varphi_n^{(1)} \\ \varphi_1^{(2)} & \dots & \varphi_n^{(2)} \\ \vdots & & \vdots \\ \varphi_1^{(n)} & & \varphi_n^{(n)} \end{bmatrix}$$

and we want

$$\boldsymbol{\Phi}=\boldsymbol{S}\boldsymbol{F}, \quad \boldsymbol{\Phi}'\boldsymbol{\Phi}=\boldsymbol{I}$$

where $\boldsymbol{S}$ is an $n \times n$ matrix. The orthogonality condition gives

$$\boldsymbol{F}'\boldsymbol{S}'\boldsymbol{F}\boldsymbol{S}=\boldsymbol{I}.$$

By linear independence $\boldsymbol{F}$ is non-singular and so is $\boldsymbol{F}'$. Hence

$$\boldsymbol{S}'\boldsymbol{S}=(\boldsymbol{F}\boldsymbol{F}')^{-1}.$$

Further $(\boldsymbol{F}\boldsymbol{F}')^{-1}$ is symmetric and we can therefore decompose it as

$$(\boldsymbol{F}\boldsymbol{F}')^{-1}=\boldsymbol{L}'\boldsymbol{L}$$

and therefore $\boldsymbol{S}$ can be chosen as the lower triangular matrix $\boldsymbol{L}$.

It is clear that this scheme produces a factorization of (any non-singular) $\boldsymbol{F}$ in the form

$$\boldsymbol{F}=\boldsymbol{L}_1\boldsymbol{\Phi}$$

where $\boldsymbol{L}_1=\boldsymbol{L}^{-1}$ is lower triangular and $\boldsymbol{\Phi}$ orthogonal. In a similar way we can produce a factorization of the form

$$\boldsymbol{F}=\boldsymbol{\Phi}\boldsymbol{R}$$

where $\boldsymbol{R}$ is upper triangular and $\boldsymbol{\Phi}$ orthogonal.

(5) We also notice that this scheme is also applicable in rectangular cases, the proof holding almost as written. Thus if the $\boldsymbol{f}_i$ are k linearly independent vectors in $\boldsymbol{R}_n$ and $\boldsymbol{L}$ is the $k \times k$ lower triangular matrix in the Cholesky factorization

$$\boldsymbol{L}'\boldsymbol{L}=(\boldsymbol{F}\boldsymbol{F}')^{-1}$$

we have that $\boldsymbol{L}\boldsymbol{F}$ is orthogonal in the sense that

$$(\boldsymbol{L}\boldsymbol{F})'\boldsymbol{L}\boldsymbol{F}=\boldsymbol{I}$$

where $\mathbf{I}$ is the $k \times k$ unit matrix. The k column vectors of $\boldsymbol{\Phi}=\mathbf{L}\,\mathbf{F}$ form an orthonormal basis for the subspace spanned by $\boldsymbol{f}_1, \boldsymbol{f}_2, \ldots, \boldsymbol{f}_k$.

We can interpret these rectangular results in terms of factorizations of the form

$$\mathbf{F}=\mathbf{L}\boldsymbol{\Phi}, \quad \mathbf{F}=\boldsymbol{\Phi}\mathbf{R}$$

where $\mathbf{F}$, $\mathbf{L}$ are $n \times k$ matrices and $\mathbf{L}$, $\mathbf{R}$ triangular $k \times k$ matrices.

7. Solution of systems, determinants

Once Theorem 4.1 is available the solution of the system $\mathbf{A}\boldsymbol{x}=\boldsymbol{b}$ is easy. We have

$$\mathbf{L}\mathbf{U}\boldsymbol{x}=\boldsymbol{b}$$

which can be written as

$$\mathbf{U}\boldsymbol{x}=\boldsymbol{y}, \quad \mathbf{L}\boldsymbol{y}=\boldsymbol{b}.$$

So, $\boldsymbol{x}$ is obtained by solving a triangular system after obtaining the right-hand side $\boldsymbol{y}$ by the solution of another triangular system. As we have seen, the solution of such systems involve about $n^2/2$ multiplications.

It is also clear that

$$\det \mathbf{A}=\det \mathbf{L}\mathbf{U}=\det \mathbf{L} \det \mathbf{U}=\prod l_{ii} \prod u_{ii}$$

and so det $\mathbf{A}$ can be evaluated by at most $2n$ additional multiplications.

What we have to do now is to find the extent of the computation of $\mathbf{L}$, $\mathbf{U}$. We begin with the symmetric case and to clarify the process, discuss a 3×3 matrix. We assume

$$\mathbf{A}=\begin{bmatrix} 1 & 2 & 3 \\ 2 & 3 & 4 \\ 3 & 4 & 4 \end{bmatrix}=\begin{bmatrix} a & 0 & 0 \\ b & c & 0 \\ d & e & f \end{bmatrix}\begin{bmatrix} a & b & d \\ 0 & c & e \\ 0 & 0 & f \end{bmatrix}.$$

Equating elements in the first row we get

$$a^2=1, \quad ab=2, \quad ad=3$$

from which, assuming $a=1$, we obtain $b=2$, $d=3$. Next equate elements in the second row.

$$b^2+c^2=3, \quad bd+ce=4$$

giving $c=+i$ say and then $e=2i$. Finally, equating elements in the third row we get

$$d^2+e^2+f^2=4 \quad \text{giving} \quad f=+i \quad \text{say}.$$

(Remark. When $\mathbf{A}$ is positive definite then $\mathbf{L}$ is real, and conversely. This is a convenient way to test the definiteness of a symmetric matrix **A**. Cf. Problems 4.11, 4.12, 4.28.)

The formulas for the general case, proceeding from row to row, are

$$\sum_{k=1}^{i} l_{ik} l_{ik} = a_{ii}, \; l_{ii} = [a_{ii} - \sum_{k=1}^{i-1} l_{ik}^2]^{1/2}$$

and

$$\sum_{k=1}^{j} l_{ik} l_{jk} = a_{ij}, \; l_{ij} = [a_{ij} - \sum_{k=1}^{j-1} l_{ik} l_{jk}]/l_{jj}, \quad j = 1, 2, \dots, i-1.$$

The calculation of the i-th row involves about $i^2/2$ multiplications and the whole calculation about $n^3/6$ multiplications. In addition there are the n square roots.

If the **LDL′** decomposition is used, no square roots are needed and, if the calculation is organized cleverly, about $n^3/6$ multiplications are still enough.

The count in the unsymmetric case, in which about $n^3/3$ multiplications are required, is set as Problem 4.22.

We continue our discussion with operation counts for the Gaussian elimination method for the solution of $\mathbf{A}\mathbf{x} = \mathbf{b}$ and of the Gauss—Jordan variation on it. We indicate the r-th stage of the triangularization:

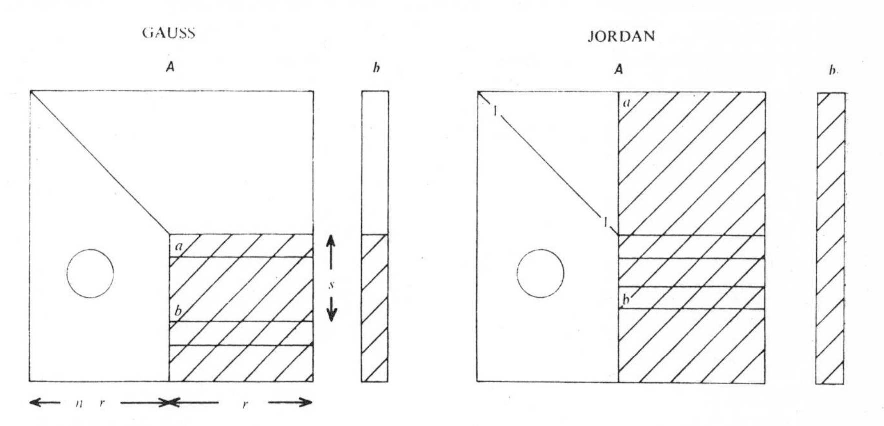

We multiply the $(n-r+1)$-st row by b/a and subtract from the $(n-r+s)$-th row. This involves 1 division, $(r-1)$ multiplications and subtractions together with a multiplication and subtraction on the right-hand side. Thus the r-th stage will involve about r^2 multiplications.

The whole triangularization therefore involves $\sum_{1}^{n} r^2 \doteqdot n^3/3$ multiplications. The back substitution, involving $\mathcal{O}(n^2)$, is negligible in comparison with this.

In the Jordan version, we need about nr multiplications at the r-th stage and therefore about $n^3/2$ multiplications altogether; it is therefore significantly more expensive then Gaussian algorithm. The determination of the x_i involves n multiplications in this case and is negligible.

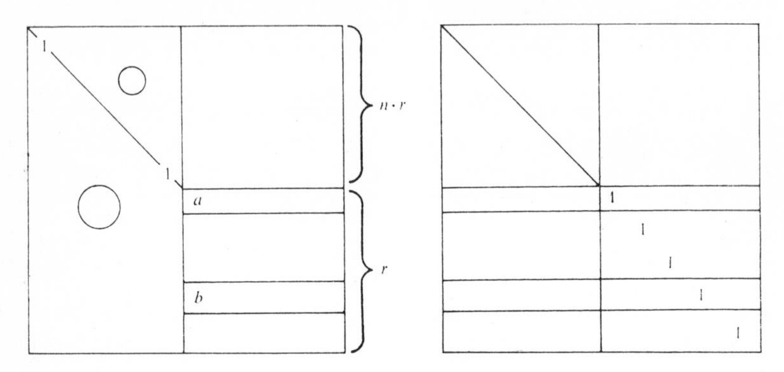

Gauss—Jordan inversion

8. Inversion of matrices

The solution of the inversion problem for a matrix $\mathbf{A}$ is clearly equivalent to the solution of n systems of linear equations:

$$[\mathbf{A}][\mathbf{A}^{-1}]=\mathbf{I}, \quad \text{i.e.,} \quad \mathbf{A}[c_1, c_2, \dots, c_n]=[e_1, e_2, \dots, e_n]$$

$$\mathbf{A}c_i=e_i, \quad i=1, 2, \dots, n$$

and is therefore certainly soluble using at most $n\times n^3/3$ multiplications, if we use the Gauss process or $n\times n^3/2$ if we use the Jordan variant. However, if we make use of the fact that it is only the right-hand sides which change and that they are indeed very special, we shall see that in either case we can get by with about n^3 multiplications.

It is best to visualize the process as one of operating on $\mathbf{A}$ and $\mathbf{I}$ instead of one on $\mathbf{A}$ and $\boldsymbol{b}$ as in the last diagram. The case of triangularization or diagonalization remains the same, and we have only to count the manipulations of the n-right hand sides and the back substitutions. We discuss the Jordan variant only.

In the r-th stage, as illustrated, we normalize the $(n-r+1)$-th row by division through by a and then subtract an appropriate multiple of it from the s-th row, for $s=1, 2, \dots, s-1, s+1, \dots, n$. In all we have about

$$n(n-r)$$

multiplications at this stage. Summing we have about

$$\sum n(n-r)=n^2\sum 1-n\sum r\doteqdot n^3-n^3/2=n^3/2$$

multiplications altogether.

The total required for inversion by the Jordan scheme is about n^3.

9. Optimality

The question as to whether cheaper ways evaluating det $\mathbf{A}$ or of solving $\mathbf{A}x=b$ or of inverting a matrix $\mathbf{A}$ exist is a natural one. A negative solution to such a question has been given by Klyuyev and Kokovkin—Shcherbak. Provided we use full row or full column operations the Gaussian scheme is optimal. However Strassen (cf. Problem 4.7) has shown by partitioning that the three basic computations can be carried out in $\mathcal{O}(n^{\log_2 7})$ operations. Winograd has shown that about $\frac{1}{2}n^3$ operations suffice and that Strassen's basic result is optimal.

Chapter 4, Problems

4.1. Show that the total number of multiplications required by the Gaussian scheme for inverting an $n\times n$ matrix $\mathbf{A}$ is approximately n^3.

4.2. If $\mathbf{A}$ is a symmetric band matrix of width $2m+1$ show that the matrix $\mathbf{L}$ occurring in the Cholesky factorization $\mathbf{A}=\mathbf{L}\mathbf{L}'$ is also a band matrix.

Let $\mathbf{B}$ be a band matrix of width b, i.e.,

$$b_{ij}=0 \quad \text{if} \quad |i-j|>b.$$

Let $\mathbf{C}$ be another band matrix of width c.

Show that $\mathbf{B}+\mathbf{C}$ and $\mathbf{BC}$ are band matrices and find their widths.

If $\mathbf{B}$ is non-singular, is $\mathbf{B}^{-1}$ a band matrix?

4.3. Express the "pivots", i.e., the leading elements in the successive Gauss transforms of $\mathbf{A}$, in terms of the leading minors of $\mathbf{A}$, i.e.

$$a_{11},\ \det\begin{bmatrix} a_{11} & a_{12} \\ a_{21} & a_{22} \end{bmatrix},\ \ldots,\ \det \mathbf{A}.$$

4.4. Show that $\mathbf{LU}$ factorization of a matrix $\mathbf{A}$ can be carried out in the following two cases:

(a) if $\mathbf{A}$ has a strictly dominant diagonal,

(b) if $\mathbf{A}$ is positive definite.

4.5. Estimate the number of operations involved in the application of the Gram—Schmidt orthogonalization process.

4.6. Express the matrices $\mathbf{A}=\begin{bmatrix}2 & -4 & 2\\ 1 & 0 & -4\\ 3 & 0 & 0\end{bmatrix}$, $\mathbf{B}=\begin{bmatrix}-7 & -30 & -65\\ 4 & 12 & -31\\ 4 & -24 & -22\end{bmatrix}$

in the form $\mathbf{QR}$, where $\mathbf{Q}$ is orthogonal and $\mathbf{R}$ upper triangular.

4.7. (Strassen—Gastinel.) Suppose $\mathbf{A}$, $\mathbf{B}$ are 2×2 matrices and that $\mathbf{C}=\mathbf{AB}$. Show that

$$\begin{bmatrix}0 & 1 & 1 & -1 & 0 & 1 & 0\\ 1 & 0 & 0 & 1 & 0 & 0 & 0\\ 0 & 1 & 0 & 0 & 1 & 0 & 0\\ 1 & 0 & 1 & 0 & -1 & 0 & 1\end{bmatrix}\begin{bmatrix}a_{11}\,(b_{12}-b_{22})\\ a_{22}\,(b_{21}-b_{11})\\ (a_{11}+a_{22})(b_{11}+b_{22})\\ (a_{12}+a_{11})\,b_{22}\\ (a_{21}+a_{22})\,b_{11}\\ (a_{12}-a_{22})(b_{21}+b_{22})\\ (a_{21}-a_{11})(b_{12}+b_{11})\end{bmatrix}=\begin{bmatrix}c_{11}\\ c_{12}\\ c_{21}\\ c_{22}\end{bmatrix}.$$

4.8. If $\mathbf{A}=\begin{bmatrix}\mathbf{A}_{11} & \mathbf{A}_{12}\\ \mathbf{A}_{21} & \mathbf{A}_{22}\end{bmatrix}$, where $\mathbf{A}_{11}$, $\mathbf{A}_{22}$ are square matrices and $\mathbf{A}_{11}$ and $\mathbf{A}_{22}-\mathbf{A}_{21}\mathbf{A}_{11}^{-1}\mathbf{A}_{12}=\mathbf{C}$ are non-singular, compute $\mathbf{AB}$ where

$$\mathbf{B}=\begin{bmatrix}\mathbf{A}_{11}^{-1}+\mathbf{A}_{11}^{-1}\mathbf{A}_{12}\mathbf{C}^{-1}\mathbf{A}_{21}\mathbf{A}_{11}^{-1}, & -\mathbf{A}_{11}^{-1}\mathbf{A}_{12}\mathbf{C}^{-1}\\ -\mathbf{C}^{-1}\mathbf{A}_{21}\mathbf{A}_{11}^{-1}, & \mathbf{C}^{-1}\end{bmatrix}.$$

4.9. Calculate $[\mathbf{X}+i\mathbf{Y}][\mathbf{Z}^{-1}-i\mathbf{Z}^{-1}\mathbf{Y}\mathbf{X}^{-1}]$ and

$$\begin{bmatrix}\mathbf{X} & \mathbf{Y}\\ -\mathbf{Y} & \mathbf{X}\end{bmatrix}\begin{bmatrix}\mathbf{Z}^{-1} & -\mathbf{Z}^{-1}\mathbf{Y}\mathbf{X}^{-1}\\ \mathbf{Z}^{-1}\mathbf{Y}\mathbf{X}^{-1} & \mathbf{Z}^{-1}\end{bmatrix}$$

where $\mathbf{X}$, $\mathbf{Y}$ are real $n\times n$ matrices and $\mathbf{X}$ and $\mathbf{Z}=(\mathbf{X}+\mathbf{Y}\mathbf{X}^{-1}\mathbf{Y})$ are non-singular.

Can you foresee any use of these results?

Discuss the case

$$\mathbf{X}=\begin{bmatrix}1 & 0\\ 0 & 1\end{bmatrix},\quad \mathbf{Y}=\begin{bmatrix}1 & 1\\ -2 & -1\end{bmatrix}.$$

4.10. If $\mathbf{A}$, $\mathbf{B}$, $\mathbf{C}$, $\mathbf{D}$ are $n\times n$ matrices and $\mathbf{A}$ is non-singular show that

$$\det\begin{bmatrix}\mathbf{A} & \mathbf{B}\\ \mathbf{C} & \mathbf{D}\end{bmatrix}=\det[-\mathbf{ACA}^{-1}\mathbf{B}+\mathbf{AD}]=\det\mathbf{A}\det[-\mathbf{CA}^{-1}\mathbf{B}+\mathbf{D}].$$

4.11. Is the matrix $\mathbf{A}=\begin{bmatrix}81 & -36 & 27 & -18\\ -36 & 116 & -62 & 68\\ 27 & -62 & 98 & -44\\ -18 & 68 & -44 & 90\end{bmatrix}$ positive definite?

Solve the system of equations $\mathbf{A}x=[252, 148, 76, 134]'$. Find $\mathbf{A}^{-1}$.

4.12. Is the matrix $\mathbf{A}=\begin{bmatrix} 1 & 2 & 3 & 4 \\ 2 & 3 & 4 & 5 \\ 3 & 4 & 4 & 5 \\ 4 & 5 & 5 & 7 \end{bmatrix}$ positive definite?

Solve the system of equations $\mathbf{A}x=[30, 40, 43, 57]'$. Find $\mathbf{A}^{-1}$.

4.13. If $\mathbf{D}=\operatorname{diag}[d_1, d_2, \ldots, d_n]$ is non-singular, find the (i, j) element in $\mathbf{D}^{-1}\mathbf{A}\mathbf{D}$. Write a program to compute $\mathbf{D}^{-1}\mathbf{A}\mathbf{D}$.

Describe in words the effect of (1) premultiplication and (2) postmultiplication by a diagonal matrix.

If $\mathbf{E}=\operatorname{diag}[1, -1, \ldots, (-1)^{n-1}]$ describe the matrix $\mathbf{E}\mathbf{A}\mathbf{E}$.

Let $\mathbf{A}$ be a triple diagonal matrix with $a_{ij}\neq 0$ for $|i-j|=1$. Show that there is a diagonal matrix $\mathbf{D}$ such that $\mathbf{D}^{-1}\mathbf{A}\mathbf{D}$ is symmetric.

4.14. Write a program to compute the determinant of a triple diagonal matrix by the method of p. 30 and apply it to $\mathbf{A}_8$, $\mathbf{A}_9^{-1}$.

4.15. Write a program to solve a triple diagonal system and apply it to cases when the matrix of the system is $\mathbf{A}_8$, $\mathbf{A}_9^{-1}$.

4.16 Determine the number of operations required to find the inverse of a triangular matrix.

4.17. Discuss the inversion of a triple diagonal matrix.

4.18. (Dekker.) Find the inverse of

$$\mathbf{M}_4=\begin{bmatrix} 4 & 2 & 4 & 1 \\ 30 & 20 & 45 & 12 \\ 20 & 15 & 36 & 10 \\ 35 & 28 & 70 & 20 \end{bmatrix}, \quad \mathbf{M}_5=\begin{bmatrix} 5 & 5 & 5 & 5 & 3 \\ 30 & 40 & 45 & 48 & 30 \\ 70 & 105 & 126 & 140 & 90 \\ 70 & 112 & 140 & 160 & 105 \\ 42 & 70 & 90 & 105 & 70 \end{bmatrix}.$$

Can you find other (non-trivial) matrices for which the inverse is obtainable in the same way as $\mathbf{M}_r^{-1}$ is obtained from $\mathbf{M}_r$?

4.19. Show how to evaluate $\det \mathbf{A}$ when $\mathbf{A}$ is in Hessenberg form, i.e., with zeros in positions (i, j) where $i=1, 2, \ldots, n-1$, $j=i+2, i+3, \ldots, n$.

Evaluate, in particular, $\det \mathbf{A}$ when $\mathbf{A}$ is

$$\begin{bmatrix} 2 & 1 & & & \\ -3 & 2 & 1 & & \\ 4 & 3 & 2 & 1 & \\ -5 & 4 & 3 & 2 & 1 \\ 6 & 5 & 4 & 3 & 2 \end{bmatrix}.$$

4.20. Discuss the inversion of a Toeplitz matrix, i.e., a matrix $\mathbf{A}=[a_{ij}]$ where the elements in the main diagonal and in all parallel ones are constant, thus:

$$\begin{bmatrix} a_0 & a_1 & a_2 \\ a_{-1} & a_0 & a_1 \\ a_{-2} & a_{-1} & a_0 \end{bmatrix}.$$

4.21. Find an upper triangular matrix $\mathbf{U}$ such that

$$\mathbf{U}'\mathbf{U}=\mathbf{A}=\begin{bmatrix} 1 & \frac{1}{2} & \frac{1}{3} & \frac{1}{4} \\ \frac{1}{2} & 1 & \frac{2}{3} & \frac{1}{2} \\ \frac{1}{3} & \frac{2}{3} & 1 & \frac{3}{4} \\ \frac{1}{4} & \frac{1}{2} & \frac{3}{4} & 1 \end{bmatrix}.$$

Find $\mathbf{U}^{-1}$ and hence find $\mathbf{A}^{-1}=\mathbf{U}^{-1}(\mathbf{U}^{-1})'$.

4.22. Find the number of operations required in the triangular decomposition of a *general* matrix $\mathbf{A}=\mathbf{LU}$.

Find the number of operations required to compute the product $\mathbf{LU}$ of two *opposite* triangular matrices.

Use these results and that of Problem 4.16 to find the cost of determining

$$\mathbf{A}^{-1}=\mathbf{U}^{-1}\mathbf{L}^{-1}.$$

4.23. (K. W. Schmidt—I. Olkin.) Let $\boldsymbol{\alpha}_k$, $\boldsymbol{\beta}_k$ be similarly positioned $k\times k$ submatrices of the matrices $\mathbf{A}$, $(\mathbf{A}^{-1})'$. Let $\mathbf{C}$ be the matrix obtained from $\mathbf{A}$ by replacing $\boldsymbol{\alpha}_k$ by $\mathbf{X}_k$. Show that

$$\det \mathbf{C}=\det \mathbf{A}\cdot \det\{\mathbf{I}_k+(\mathbf{X}_k-\boldsymbol{\alpha}_k)(\boldsymbol{\beta}_k)'\}.$$

4.24. Show that if $\boldsymbol{x}=[x_1, \hat{x}]'$ is a (complex) vector of length 1, i.e. $\boldsymbol{x}^*\boldsymbol{x}=1$, and if $x_1\neq -1$ is real then the matrix

$$\mathbf{U}=\begin{bmatrix} x_1 & \hat{\boldsymbol{x}}^* \\ \hat{\boldsymbol{x}} & -\mathbf{I}+(1+x_1)^{-1}\hat{\boldsymbol{x}}\hat{\boldsymbol{x}}^* \end{bmatrix},$$

which has $\boldsymbol{x}$ as its first column, is unitary.

4.25. Describe how to obtain the reduced row-echelon form of a matrix.

What difficulties can arise if one attempts to implement this reduction algorithm on a computer?

4.26. Suppose the n-dimensional column vectors $\boldsymbol{x}_1, \boldsymbol{x}_2, \ldots, \boldsymbol{x}_r$ are linearly independent. Describe how to obtain $n-r$ vectors $\boldsymbol{x}_{r+1}, \boldsymbol{x}_{r+2}, \ldots, \boldsymbol{x}_n$ so that $\boldsymbol{x}_1, \ldots, \boldsymbol{x}_n$ forms a basis for $\mathbf{R}_n$.

Discuss the implementation of this process on a computer.

4.27. Find the inverse of the block matrix

$$\mathcal{A}=\begin{bmatrix} \mathbf{I} & \mathbf{A} & 0 \\ 0 & \mathbf{I} & \mathbf{B} \\ 0 & 0 & \mathbf{I} \end{bmatrix}.$$

4.28. Confirm that the matrix $\mathbf{A}$ of p. 36 is not positive definite:

(1) Show that not all its characteristic values are positive.

(2) Show that there is a vector $\boldsymbol{x}$ such that the quadratic form $\boldsymbol{x}'\mathbf{A}\,\boldsymbol{x}$ is negative.

(3) Show that not all the leading principal minors of $\mathbf{A}$ are positive.

CHAPTER 5

The Inversion Problem II: Practical Computation

We now return to the study of our problems from the point of view "practical computation" as distinct from that of "theoretical arithmetic" which we took in Chapter 4. Several questions clearly arise as is indicated by Problems 5.1, 5.2, 5.3, 5.6, which are to be solved in the first place by hand or by desk machine. We discuss these briefly.

Problem 5.1 indicates that a small perturbation of a single element in a matrix *may* cause enormous changes in its determinant. Problem 5.2 shows that a system of linear equations can have the property that a "solution vector" widely different from the true solution can have very small residuals. Problem 5.3 shows that small perturbations in the right-hand sides of a system of linear equations can cause large changes in the true solutions. Problem 5.6 shows that a matrix can have a very good approximate right inverse which is a very bad approximate left inverse.

These phenomena have been exhibited in the case of matrices of orders 2, 3, 4. It is to be expected that more violent things can occur when we are dealing with matrices of order 100 or more which are common in computational practice, and where the forced perturbations in our examples are replaced by those caused by the round-offs in the computation.

Several questions have to be discussed. First, can we readily find in advance whether a matrix on which we are working is or is not peculiarly sensitive — the technical word used is "ill-conditioned"? Second, we would like to know whether this sensitivity is an intrinsic property of the matrix, and is in evidence no matter what numerical method we use. Third, we want to know how to deal with bad cases. This problem is discussed in Chapter 9.

With regard to the first question a considerable body of theory exists, and this is supported by much computational evidence, which indicates that the bad behavior or "*ill-condition*" of a matrix $\boldsymbol{A}$ depends quantitatively on its condition number.

$$\varkappa(\boldsymbol{A}) = \|\boldsymbol{A}\|\,\|\boldsymbol{A}^{-1}\|.$$

If this is large we may expect trouble. Here are some representative values in the case of 4×4 matrices

$$\varkappa_\infty(\boldsymbol{C}_4) = 12, \quad \varkappa_\infty(\boldsymbol{W}) = 4{,}488, \quad \varkappa_\infty(\boldsymbol{H}_4) = 28{,}375, \quad \varkappa_\infty(\boldsymbol{R}) = 62{,}608$$

when the norm used is the maximum absolute row sum. The troubles which occur in the second and third and fourth cases are illustrated in Problems 5.3, 5.4, 5.5, 5.9.

The P-condition number, i.e., that associated with the matrix norm induced by the euclidean vector norm, has been calculated for various matrices. For instance we have, where α is a positive constant,

$$\varkappa_2(\mathbf{I})=1, \quad \varkappa_2(\mathbf{C}_n)=\mathcal{O}(n^2), \quad \varkappa_2(\mathbf{H}_n)=\mathcal{O}(\exp \alpha n)$$

and the behavior of the matrices with respect to inversion is correlated with their condition number.

We now indicate some theoretical evidence supporting our contention that $\varkappa(\mathbf{A})$ is related to the sensitivity of the linear equation problem.

Suppose ξ is the perturbation in the solution $\boldsymbol{x}$ of the non-singular system

$$\mathbf{A}\boldsymbol{x}=\boldsymbol{b}$$

caused by a perturbation $\boldsymbol{\beta}$ in $\boldsymbol{b}$, i.e.,

$$\mathbf{A}(\boldsymbol{x}+\xi)=\boldsymbol{b}+\boldsymbol{\beta}.$$

The relative error in the solution is $\|\xi\|/\|\boldsymbol{x}\|$ while that in the data is $\|\boldsymbol{\beta}\|/\|\boldsymbol{b}\|$, where $\|\cdot\|$ indicates any vector norm. We have

$$\mathbf{A}\boldsymbol{x}=\boldsymbol{b} \quad \text{and so} \quad \|\boldsymbol{b}\| \leqq \|\mathbf{A}\|\,\|\boldsymbol{x}\|, \quad \text{equality for some } \boldsymbol{b}$$

and

$$\mathbf{A}\xi=\boldsymbol{\beta} \quad \text{so that} \quad \xi=\mathbf{A}^{-1}\boldsymbol{\beta} \quad \text{and} \quad \|\xi\| \leqq \|\mathbf{A}^{-1}\|\,\|\boldsymbol{\beta}\|, \quad \text{equality for some } \boldsymbol{\beta},$$

where the matrix norm is that induced by the chosen vector norm. Clearly

$$\frac{\|\xi\|}{\|\boldsymbol{x}\|}\Big/\frac{\|\boldsymbol{\beta}\|}{\|\boldsymbol{b}\|}=\frac{\|\xi\|}{\|\boldsymbol{\beta}\|}\cdot\frac{\|\boldsymbol{b}\|}{\|\boldsymbol{x}\|} \leqq \|\mathbf{A}^{-1}\|\,\|\mathbf{A}\|=\varkappa(\mathbf{A}), \quad \text{equality for above } \boldsymbol{b}, \boldsymbol{\beta}.$$

Similar results can be obtained to show the effect of a perturbation on $\mathbf{A}$ on the solution or on the inverse. See Problem 5.20.

On the whole, the Gaussian elimination process, or some variant on it, is the most satisfactory one for the solution of a general system of linear equations — for "sparse" systems, the iterative methods which are discussed later, are often better. From the point of view of speed see the remarks on p. 39. Certain changes have to be made in the classical algorithm from the point of view of 'practical computation'. The system

$$\left.\begin{array}{l} 0\cdot x+1\cdot y=1 \\ 1\cdot x+0\cdot y=2 \end{array}\right\}$$

falls outside the scope of our algorithms since $a_{11}=0$, until we interchange the order of the equations. If we consider instead

$$\left.\begin{aligned}10^{-2}x+1\cdot y=1\\1\cdot x+1\cdot y=2\end{aligned}\right\}$$

we *can* proceed but (Problem 6.17) our results improve if we interchange the order. This process of "positioning for size" or "pivoting" is essential if we want an algorithm efficient from the point of view of practical computation for general cases. (In certain special classes of systems it can be found that this is unnecessary; see, e.g. Problem 4.4.)

If we use "complete pivoting" we begin by choosing as the pivot an element a_{ij} of maximum absolute value which we imagine put in the (1, 1) position. After killing the remaining elements in the first column we choose as our next pivot the element $a_{ij}^{(2)}$ of maximum absolute value among those left in the last $n-1$ rows, and this we imagine put in the (2, 2) position. And so on.

If we use "partial pivoting" we only scan the elements in the first row $a_{11}^{(1)}, a_{12}^{(1)}, \ldots, a_{1n}^{(1)}$ to select the first pivot. Then we scan the elements $a_{22}^{(2)}, a_{23}^{(2)}, \ldots$ $\ldots, a_{2n}^{(2)}$ to select the second pivot. And so on.

Before using any library subroutine, it is essential to check that such devices are incorporated. The extra time needed for possible interchanges is negligible, granted efficient programming.

In order to evaluate a library subroutine for, say, inversion, we can proceed as follows.

Applying such a program to a non-singular matrix $\mathbf{A}$ we (usually) get as an output an approximate or machine inverse of $\mathbf{A}$, which we denote by $\mathbf{X}$. We want to know how good an inverse is $\mathbf{X}$. If we know $\mathbf{A}^{-1}$ then we can compute some norm

$$\|\mathbf{A}^{-1}-\mathbf{X}\|$$

and this should be small. Alternatively we can check whether $\mathbf{X}$ is a good right or left inverse by computing

$$\|\mathbf{AX}-\mathbf{I}\| \quad \text{or} \quad \|\mathbf{XA}-\mathbf{I}\|.$$

We could also find the machine inverse $\mathcal{X}$ of $\mathbf{X}$ and compute

$$\|\mathbf{A}-\mathcal{X}\|.$$

In view of the equivalence of norms, the actual norm used is not critical.

We would like to be sure of the quality of our inverse for *all* the matrices we are likely to encounter and for this purpose we set up the battery of test matrices which are listed in Problem 1.2. We have set, as Problem 5.14, the testing of local inversion routines.

Here are some actual results for the matrix $\mathbf{C}_n$ and the matrix $\mathbf{C}_n^2$ which arises in the numerical solution of the biharmonic equation. Note that $\varkappa_2(\mathbf{C}_n^2)=$ $=\mathcal{O}(n^4)$.

	$n=10$	$n=20$	$n=30$	$n=40$	$n=50$
$\mathbf{C}_n$	2×10^{-7}	6×10^{-7}	1×10^{-6}	2×10^{-6}	3×10^{-6}
$\mathbf{C}_n^2$	3×10^{-5}	6×10^{-4}	2×10^{-4}	9×10^{-3}	2×10^{-2}.

These results were obtained by single precision calculations (8 bits characteristic, 27 bits mantissa) and the numbers tabulated are the euclidean norm $\|\mathbf{AX}-\mathbf{I}\|$ where $\mathbf{X}$ is the alleged inverse of $\mathbf{A}$. With a double precision calculation the corresponding results for $\mathbf{C}_n$ are

$\mathbf{C}_n$	3×10^{-14}	3×10^{-13}	1×10^{-12}	4×10^{-12}	6×10^{-12}

The results obtained in Problem 5.14 indicate that with a good inversion program using the Gaussian elimination method, with pivoting, the errors incurred are roughly proportional to a condition number. More precisely they appear proportional to the product of the condition number and a total round-off error, $n^2\varepsilon$ where ε is the round-off error. To establish results of this kind requires a quite elaborate analysis. The early results of von Neumann and Goldstine were of the forward kind:

$$\|\mathbf{A}^{-1}-\mathbf{X}\|\sim\varkappa(\mathbf{A})$$

while more recent results of Wilkinson are of the backward kind

$$\|\mathbf{A}-\mathbf{X}^{-1}\|\sim\varkappa(\mathbf{A})$$

where $\mathbf{X}$ is the machine inverse of $\mathbf{A}$ and $\mathbf{A}^{-1}$, $\mathbf{X}^{-1}$ are the true inverses of $\mathbf{A}$, $\mathbf{X}$.

Chapter 5, Problems

5.1. (Etherington.) Evaluate the determinants det $\mathbf{A}$, det $\mathbf{B}$, det $\mathbf{C}_4$, det $(\mathbf{C}_4+\mathbf{E})$ where

$$\mathbf{A}=\begin{bmatrix}-73 & 78 & 24\\ 92 & 66 & 25\\ -80 & 37 & 10\end{bmatrix},\quad \mathbf{B}=\begin{bmatrix}-73 & 78 & 24\\ 92 & 66 & 25\\ -80 & 37 & 10.01\end{bmatrix},$$

$$\mathbf{C}_4=\begin{bmatrix}-2 & 1 & 0 & 0\\ 1 & -2 & 1 & 0\\ 0 & 1 & -2 & 1\\ 0 & 0 & 1 & -2\end{bmatrix},\quad \mathbf{E}=\begin{bmatrix}0 & 0 & 0 & 0\\ 0 & 0 & 0 & 0\\ 0 & 0 & 0 & 0\\ 0 & 0 & 0. & 01\end{bmatrix}.$$

Evaluate also $\mathbf{A}^{-1}$, $\mathbf{C}_4^{-1}$ and the matrix norms of $\mathbf{A}$, $\mathbf{A}^{-1}$, $\mathbf{C}_4$, $\mathbf{C}_4^{-1}$ induced by the Chebyshev, Manhattan and euclidean vector norms.

Find the characteristic polynomial and the characteristic roots of $\mathbf{A}$.

5.2. (Kahan.) Solve the system $\mathbf{A}\boldsymbol{x}=\boldsymbol{b}$,

(a) exactly (i.e., by hand or desk machine)

and

(b) using the console,

where

$$\mathbf{A}=\begin{bmatrix} .2161 & .1441 \\ 1.2969 & .8648 \end{bmatrix}, \quad \boldsymbol{b}=\begin{bmatrix} .1440 \\ .8642 \end{bmatrix}.$$

Find the residual $\boldsymbol{r}=\boldsymbol{b}-\mathbf{A}\boldsymbol{z}$ when

$$\boldsymbol{z}=\begin{bmatrix} .9911 \\ -.4870 \end{bmatrix}.$$

Find the inverse of $\mathbf{A}$ and also $\|\mathbf{A}\|$, $\|\mathbf{A}^{-1}\|$ for the matrix norms induced by the Chebyshev and Manhattan vector norms.

5.3. (Wilson.) Solve the system of equations $\mathbf{W}\boldsymbol{x}=\boldsymbol{b}$ when

$\boldsymbol{b}'=[32, 23, 33, 31]$ and when $\boldsymbol{b}'=[32.01, 22.99, 33.01, 30.99]$ where

$$\mathbf{W}=\begin{bmatrix} 10 & 7 & 8 & 7 \\ 7 & 5 & 6 & 5 \\ 8 & 6 & 10 & 9 \\ 7 & 5 & 9 & 10 \end{bmatrix}.$$

5.4. Show, using the $\mathbf{L}\mathbf{L}'$ factorization, that the inverse of the matrix $\mathbf{W}$ in Problem 5.3 is

$$\mathbf{W}^{-1}=\begin{bmatrix} 25 & -41 & 10 & -6 \\ -41 & 68 & -17 & 10 \\ 10 & -17 & 5 & -3 \\ -6 & 10 & -3 & 2 \end{bmatrix}.$$

Find the matrix norms of $\mathbf{W}$, $\mathbf{W}^{-1}$ induced by the Chebyshev and Manhattan vector norms.

5.5. Express the matrix $\mathbf{H}_4=\mathbf{A}_{13}$ as a product $\mathbf{L}\mathbf{D}\mathbf{L}'$ where the diagonal elements of $\mathbf{L}$ are all 1. Also express $\mathbf{H}_4$ in the form $\mathcal{L}\mathcal{L}'$. Find $\mathbf{H}_4^{-1}$.

(All of this problem is to be done exactly in the first place and then machine results should be obtained and compared.)

Find the matrix norms of $\boldsymbol{H}_4$, $\boldsymbol{H}_4^{-1}$ induced by the Chebyshev and Manhattan vector norms.

5.6. If $\boldsymbol{A}=\begin{bmatrix}1 & 1\\ 1 & 1-n^{-2}\end{bmatrix}$, $\quad \boldsymbol{B}=\begin{bmatrix}1+n-n^2 & n^2\\ n^2-n & -n^2\end{bmatrix}$

evaluate $\boldsymbol{A}^{-1}$, $\boldsymbol{AB}-\boldsymbol{I}$, $\boldsymbol{BA}-\boldsymbol{I}$ and the Chebyshev norms of the matrices.

5.7. (Moulton.) Discuss the solution of the system of equations

$$0.34622\,x+0.35381\,y+0.36518\,z=0.24561,$$
$$0.89318\,x+0.90274\,y+0.91143\,z=0.62433,$$
$$0.22431\,x+0.23642\,y+0.24375\,z=0.17145.$$

In particular, evaluate the residuals for the following three sets of values of x, y, z:

$$\begin{array}{cccc} x: & -1.027066 & -1.022773 & -1.031229\\ y: & 2.091962 & 2.084125 & 2.099457\\ z: & -0.380515 & -0.376941 & -0.383879. \end{array}$$

5.8. (Morris-Neville.) Discuss the solution of the system of equations:

$$\begin{bmatrix} 5.39999 & 5.23286 & 4.35785 & 3.62242 & 2.76472 & 1.84691\\ 5.23286 & 7.87190 & 3.62242 & 5.25651 & 1.84691 & 2.80269\\ 4.35785 & 3.62242 & 3.88141 & 2.97304 & 2.63974 & 1.67936\\ 3.62242 & 5.25651 & 2.97304 & 4.37677 & 1.67936 & 2.63246\\ 2.76472 & 1.84691 & 2.63974 & 1.67936 & 2.01578 & 1.14921\\ 1.84691 & 2.80269 & 1.67936 & 2.63246 & 1.14921 & 1.94065 \end{bmatrix} \begin{bmatrix} x\\ y\\ z\\ u\\ v\\ w \end{bmatrix} = \begin{bmatrix} 1.23679\\ 0.48448\\ 1.24950\\ 0.47304\\ 1.06470\\ 0.37831 \end{bmatrix}.$$

5.9. (Rutishauser.) Find the inverse of the positive definite symmetric matrix

$$\boldsymbol{R}=\begin{bmatrix} 10 & 1 & 4 & 0\\ 1 & 10 & 5 & -1\\ 4 & 5 & 10 & 7\\ 0 & -1 & 7 & 9 \end{bmatrix},$$

using the machine.

Find some of the condition numbers of this matrix.

5.10. Solve exactly the system

$$\boldsymbol{A}\boldsymbol{x}=\boldsymbol{b}$$

where $\quad \boldsymbol{A}=\begin{bmatrix} 5 & 7 & 3\\ 7 & 11 & 2\\ 3 & 2 & 6 \end{bmatrix}, \quad \boldsymbol{b}=\begin{bmatrix} 0\\ -1\\ 0 \end{bmatrix}.$

Find the condition number of $\boldsymbol{A}$ with respect to the Chebyshev norm.

Solve this system using the subroutine provided.

Repeat this problem in the case of the matrix

$$\boldsymbol{C}_3=\begin{bmatrix}-2 & 1 & 0\\ 1 & -2 & 1\\ 0 & 1 & -2\end{bmatrix}.$$

5.11. Is the system

$$\begin{aligned}6x+13y-17z&=1\\ 13x+29y-38z&=2\\ -17x-38y+50z&=-3\end{aligned}$$

ill-conditioned?

5.12. Calculate $\varkappa_\infty(\boldsymbol{A})$ where

$$\boldsymbol{A}=\begin{bmatrix}1 & 1 & 1\\ 1 & 10 & 10^2\\ 1 & 10^2 & 10^4\end{bmatrix}.$$

Calculate also $\varkappa_\infty(\boldsymbol{D}\boldsymbol{A})$ where

$$\boldsymbol{D}=\operatorname{diag}[1/3,\ 1/111,\ 1/10101].$$

Do these results suggest anything?

5.13. Find the inverses and determinants of the matrices listed in Problem 1.2.

Find some condition numbers of these matrices, either exactly or approximately.

5.14. For each of the matrices above, for appropriate values of n, and for the available inversion programs, compute norms of the four error matrices. Record that for $\boldsymbol{AX}-\boldsymbol{I}$ in the Frobenius case.

[In the case of $\boldsymbol{A}_1, \boldsymbol{A}_3, \boldsymbol{A}_7, \boldsymbol{A}_8, \boldsymbol{A}_9, \boldsymbol{A}_{14}, \boldsymbol{A}_{15}$, $n=20$, $n=50$ are appropriate; for $\boldsymbol{A}_{16}$, $n=10$, $n=20$ are appropriate and for $\boldsymbol{A}_{13}$, $n=4$, $n=6$, $n=8$, $n=10$ are appropriate.]

5.15. We have seen that if $\boldsymbol{\xi}$ is the perturbation in the solution $\boldsymbol{x}$ of the equation $\boldsymbol{A}\boldsymbol{x}=\boldsymbol{b}$ caused by a perturbation $\boldsymbol{\beta}$ in the right-hand side then the amplification of the relative input error is at most $\varkappa(\boldsymbol{A})$, the condition number of $\boldsymbol{A}$ with respect to the appropriate norm. We have also noted that this amplification is sharp, i.e., it is attained for certain $\boldsymbol{b}$, $\boldsymbol{\beta}$. Examine this sharpness in the case of the matrix $\boldsymbol{W}$ (of Problem 5.3) for the three standard norms. Do the same for the matrix $\boldsymbol{C}_4$ of Problem 5.1.

5.16. It is often of interest to find the change in the inverse of a matrix $\mathbf{A}$ caused by a change of special form in $\mathbf{A}$, in particular when the perturbation is an anti-scalar product $\boldsymbol{x}\boldsymbol{y}'$.

Establish the relation

$$(\mathbf{A}+\boldsymbol{x}\boldsymbol{y}')^{-1}=\mathbf{A}^{-1}-\frac{\mathbf{A}^{-1}\boldsymbol{x}\boldsymbol{y}'\mathbf{A}^{-1}}{(1+\boldsymbol{y}'\mathbf{A}^{-1}\boldsymbol{x})}$$

assuming $1+\boldsymbol{y}'\mathbf{A}^{-1}\boldsymbol{x}\neq 0$.

Apply this in a special case, e.g.

$$\mathbf{A}=\mathbf{C}_4=\begin{bmatrix} 2 & -1 & 0 & 0 \\ -1 & 2 & -1 & 0 \\ 0 & -1 & 2 & -1 \\ 0 & 0 & -1 & 2 \end{bmatrix}, \quad \boldsymbol{x}=\boldsymbol{y}=[1,\ 1,\ 1,\ 1]'.$$

Estimate the number of multiplications/divisions needed when $\mathbf{A}$ is of order n, n large.

5.17. Solve the system of equations

$$\left.\begin{aligned} 10^{-2}x+y&=1 \\ x+y&=2 \end{aligned}\right\}$$

(a) exactly, (b) by Gaussian elimination *simulating* two-decimal floating arithmetic, and (c) as in (b) but interchanging the equations.

5.18. Solve the first system of equations in Problem 5.3, simulating two-decimal arithmetic, using "complete pivoting".

5.19. (Gastinel.) Show that when $\mathbf{A}$ is non-singular and the matrix norm is induced by a p-vector norm

$$\min_{\mathbf{A}+\mathbf{F}\text{ singular}} \|\mathbf{F}\|=\frac{1}{\|\mathbf{A}^{-1}\|}.$$

Interpret this result.

5.20. (a) If $\|\mathbf{A}^{-1}\mathbf{F}\|<1$ and $\boldsymbol{\xi}$ is the perturbation of the solution of $\mathbf{A}\boldsymbol{x}=\boldsymbol{b}$ caused by a perturbation $\mathbf{F}$ in $\mathbf{A}$, i.e.

$$(\mathbf{A}+\mathbf{F})(\boldsymbol{x}+\boldsymbol{\xi})=\boldsymbol{b}$$

show that

$$\frac{\|\boldsymbol{\xi}\|/\|\boldsymbol{x}\|}{\|\mathbf{F}\|/\|\mathbf{A}\|}\leq\frac{\|\mathbf{A}\|\,\|\mathbf{A}^{-1}\|}{\{1-\|\mathbf{A}\|\,\|\mathbf{A}^{-1}\|(\|\mathbf{F}\|/\|\mathbf{A}\|)\}}.$$

(b) If we write $\mathbf{A}+\mathbf{F}=\mathbf{B}$ show that

$$\frac{\|\mathbf{A}^{-1}-\mathbf{B}^{-1}\|}{\|\mathbf{A}^{-1}\|} \leqq \|\mathbf{A}\| \, \|\mathbf{A}^{-1}\| \frac{\|\mathbf{F}\|}{\|\mathbf{A}+\mathbf{F}\|}.$$

5.21. Denote by $\mathbf{B}=\mathbf{B}_n$ the $n\times n$ lower triangular matrix having for its successive rows the coefficients in the binomial expansion $(1-1)^r$ so that

$$\mathbf{B}_4=\begin{bmatrix} 1 & 0 & 0 & 0 \\ 1 & -1 & 0 & 0 \\ 1 & -2 & 1 & 0 \\ 1 & -3 & 3 & -1 \end{bmatrix};$$

show that $\mathbf{B}_n^2=\mathbf{I}$.

Evaluate the elements of $\mathbf{A}_n=\mathbf{B}_n\mathbf{B}_n'$ and experiment with the inversion of $\mathbf{A}_n$.

5.22. (a) (Tee.) Discuss the condition of the matrix

$$\mathbf{A}=\begin{bmatrix} 11 & 10 & 14 \\ 12 & 11 & -13 \\ 14 & 13 & -66 \end{bmatrix}.$$

(b) (Zielke.) Discuss the condition of the matrix

$$\mathbf{Z}=\begin{bmatrix} 11+\alpha & 10+\alpha & 14+\alpha \\ 12+\alpha & 11+\alpha & -13+\alpha \\ 14+\alpha & 13+\alpha & -66+\alpha \end{bmatrix}.$$

5.23. (Taussky.) Discuss the relative size of the condition numbers of $\mathbf{A}$ and $\mathbf{A}\mathbf{A}'$.

CHAPTER 6

The Characteristic Value Problem—Generalities

It has been said that the main problem of applied mathematics is to find the characteristic values and vectors of symmetric matrices, or, what is the same thing, the principal axes of ellipsoids. In some cases we only require crude estimates for the characteristic values, in others we require close approximations to a single characteristic value — very often the dominant one — or, may be, to a few, while sometimes we may require all the characteristic values and vectors. The problem becomes very much more difficult if the matrix is not symmetric — in the first place, the characteristic values are no longer necessarily real. We also encounter the so-called "generalized characteristic value problem", the determination of non-trivial solutions of

$$\boldsymbol{A}\boldsymbol{x}=\lambda\boldsymbol{B}\boldsymbol{x}.$$

In this chapter we discuss or review some relevant results from the theory of matrices, assuming some familiarity with the theory of characteristic values and vectors. We also find it convenient to discuss quadratic forms, particularly positive definite forms.

1. Row and column characteristic vectors

We observe that we can define both left (or row) and right (or column) characteristic values and vectors:

$\boldsymbol{A}\boldsymbol{x}=\lambda\boldsymbol{x}$ according to our convention that $\boldsymbol{x}$ is a column vector leads to the column case while $\boldsymbol{x}'\boldsymbol{A}=\mu\boldsymbol{x}'$ leads to the row case.

For instance

$$\begin{bmatrix}7 & -10\\ 3 & -4\end{bmatrix}\begin{bmatrix}x\\ y\end{bmatrix}=\alpha\begin{bmatrix}x\\ y\end{bmatrix}$$

for

$$\alpha=2,\quad \begin{bmatrix}x\\ y\end{bmatrix}=\begin{bmatrix}2\\ 1\end{bmatrix}\quad\text{and}\quad \alpha=1,\quad \begin{bmatrix}x\\ y\end{bmatrix}=\begin{bmatrix}5\\ 3\end{bmatrix}$$

while

$$[x, y]\begin{bmatrix}7 & -10\\ 3 & -4\end{bmatrix}=\alpha[x, y]$$

for

$$\alpha=2,\quad [x, y]=[3, -5]\quad\text{and}\quad \alpha=1,\quad [x, y]=[1, -2].$$

Observe that the left and right characteristic *values* are always the same (since $\det(\mathbf{A}-\lambda\mathbf{I})=\det(\mathbf{A}'-\lambda\mathbf{I})$) but the left and right characteristic vectors may be different. However if $\mathbf{A}$ is symmetric the left and right characteristic vectors are the same.

2. Localization of characteristic values

The simplest tool for estimating the position of the characteristic values of a matrix is

(6.1) Gerschgorin's Theorem. *If* $\mathbf{A}$ *is any matrix denote by* $\alpha_1, \dots, \alpha_n$ *its characteristic values. Denote by* Γ_i *the i-th Gerschgorin disk*

$$\Gamma_i : |z-a_{ii}| \leqq \Lambda_i = \sum_{\substack{j=1 \\ j\neq i}}^{n} |a_{ij}|.$$

Denote by Ω *the union of the n disks* Γ_i. *Then all the* α*'s lie in* Ω.

There is a similar result for "columns": the j-th "column-disk" has radius $\sum_{\substack{i=1 \\ i\neq j}}^{n} |a_{ij}|$. If Ω_1 is the union of the n column-disks then all the characteristic values of $\mathbf{A}$ lie in the intersection of Ω and Ω_1.

This result is equivalent to the

(6.2) Dominant Diagonal Theorem. *A matrix* $\mathbf{A}$ *is said to have a strictly dominant diagonal if*

$$|a_{ii}| > \Lambda_i, \quad i=1, 2, \dots, n.$$

Such a matrix is non-singular.

We prove first the equivalence of these two theorems.

Assuming the Gerschgorin theorem we see that the set Ω, for a strictly dominant diagonal matrix, cannot contain the origin. Hence no characteristic value of $\mathbf{A}$ vanishes, i.e., $\mathbf{A}$ is non-singular.

Conversely, assume the dominant diagonal theorem. The matrix $\mathbf{A}-z\mathbf{I}$, where $z \notin \Omega$, has a strictly dominant diagonal and so is non-singular, i.e., any $z \notin \Omega$ cannot be a characteristic value of $\mathbf{A}$ which is to say that all the characteristic values α belong to Ω.

We shall now establish the second theorem. Assume that $\mathbf{A}$ has a strictly dominant diagonal and that it is singular. Then the homogeneous system

$$\mathbf{A}\mathbf{x}=0$$

has a non-trivial solution $\mathbf{x}=[x_1, \dots, x_n]'$. Then $\max |x_i| \neq 0$ and we may assume that $\max |x_i| = x_k = 1$. Consider the k-th equation

$$a_{kk}x_k = -\sum_j{}' a_{kj}x_j.$$

This gives, since $x_k=1$,

$$
\begin{aligned}
|a_{kk}| &\leqq \sum_j{}' |a_{kj}|\,|x_j| \\
&\leqq \sum_j{}' |a_{kj}| \quad \text{since} \quad |x_j| \leqq x_k = 1 \\
&= \Lambda_k
\end{aligned}
$$

which contradicts the assumption that the diagonal is strictly dominant.

There are many improvements and generalizations of these two theorems which, e.g., allow for the possibility of some but not all of the relations

$$|a_{ii}| > \Lambda_i$$

being equalities. For instance they allow us to conclude that $\mathbf{A}_8$ is non-singular although most of the relations $|a_{ii}| \geqq \Lambda_i$ are equalities. Some of the changes to be made can be motivated by observing some special matrices, e.g.,

$$\begin{bmatrix} 0 & 0 \\ 0 & 0 \end{bmatrix}, \quad \begin{bmatrix} 1 & 0 \\ 0 & 0 \end{bmatrix}, \quad \begin{bmatrix} 1 & 1 \\ 1 & 0 \end{bmatrix}.$$

There are many other proofs of the Gerschgorin Theorem. (See O. Taussky, Amer. Math. Monthly *56,* 672—676, (1939) and Problem 6.7.)

3. Positive matrices, non-negative matrices

In many parts of numerical analysis matrices all of whose elements are non-negative are of importance. Such matrices arise e.g., in mathematical economics. There is available an elegant theory of *non-negative* matrices: for simplicity we state the basic theorem for *positive* matrices.

(6.3) Perron-Frobenius Theorem. *If $\mathbf{A}=[a_{ij}]$ has $a_{ij}>0$ for all i, j then there is positive characteristic value of $\mathbf{A}$ which is dominant, i.e., if $\varrho(\mathbf{A})=\alpha$ then α is a characteristic value of $\mathbf{A}$ and if $\alpha' \neq \alpha$ is also a characteristic value of $\mathbf{A}$ then $\alpha > |\alpha'|$ so that, in particular, α is simple. Further the characteristic vector of $\mathbf{A}$ belonging to α can be chosen to have all its components positive: Indeed no other characteristic vector of $\mathbf{A}$ can have its components all non-negative.*

Some idea of the changes in the theorem when we change the hypothesis to non-negativity can be obtained by observing the matrices

$$\begin{bmatrix} 1 & 0 \\ 0 & 1 \end{bmatrix}, \quad \begin{bmatrix} 0 & 1 \\ 1 & 0 \end{bmatrix}, \quad \begin{bmatrix} 1 & 1 \\ 0 & 1 \end{bmatrix}.$$

We sketch a proof of the two-dimensional case of this theorem by a fixed-point method. We begin by noting that if f is a continuous function on

[0, 1] with values in [0, 1], then there is an x, $0 \leqq x \leqq 1$, such that $f(x)=x$, i.e., the transformation $x \to f(x)$ has a fixed point. See diagram.

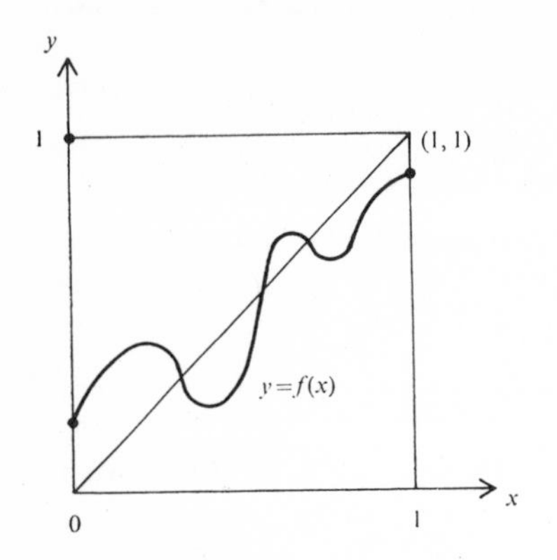

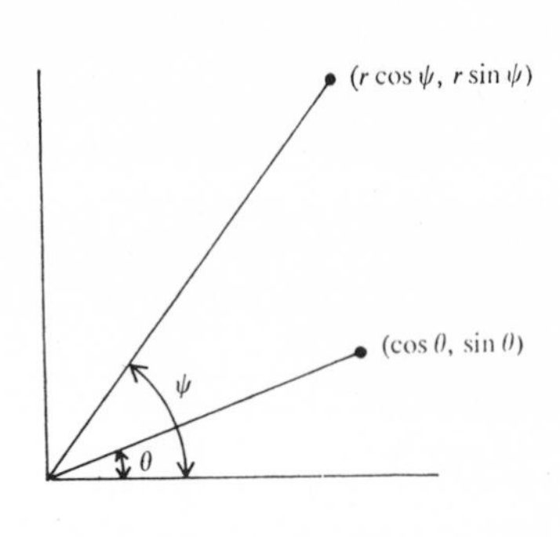

Let $\mathbf{A}$ be a 2×2 positive matrix. Then if we take an angle θ, $0 \leqq \theta \leqq \pi/2$ we can construct an angle $\psi = f(\theta)$, according to

$$\mathbf{A}\begin{bmatrix}\cos\theta\\ \sin\theta\end{bmatrix} = \begin{bmatrix}a_{11}\cos\theta + a_{12}\sin\theta\\ a_{21}\cos\theta + a_{22}\sin\theta\end{bmatrix} = \begin{bmatrix}r\cos\psi\\ r\sin\psi\end{bmatrix} = r\begin{bmatrix}\cos\psi\\ \sin\psi\end{bmatrix}$$

and then $0 \leqq \psi \leqq \pi/2$. According to the remark in the preceding paragraph the mapping $\theta \to \psi$ has a fixed point, i.e., a fixed direction, i.e. a characteristic vector which is positive. See diagram above.

4. Quadratic forms

With a real symmetric $n \times n$ matrix we associate a quadratic form

$$\mathbf{x}'\mathbf{A}\mathbf{x} = \sum_{i,\, j=1}^{n} a_{ij} x_i x_j .$$

Conversely, with a quadratic form

$$Q = \sum_i q_{ii} x_i^2 + 2 \sum_{\substack{i,\, j=1\\ i<j}}^{n} q_{ij} x_i x_j$$

we associate a symmetric matrix $\mathbf{Q}$. [In the same way with a (complex) hermitian matrix we associate an hermitian form $\mathbf{x}^* \mathbf{A} \mathbf{x}$ and conversely. The theory of hermitian forms can be developed in a fashion parallel to that for quadratic forms.]

Def. 1. *$\mathbf{Q}$ or $\mathbf{A}$ is said to be positive definite if $\mathbf{x}'\mathbf{A}\mathbf{x} \geqq 0$ for all (real) $\mathbf{x}$ and if $\mathbf{x}'\mathbf{A}\mathbf{x}=0$ only if $\mathbf{x}=0$.*

Formal definitions of negative definite and of (positive and negative) semidefinite can be given. Regarded as quadratic forms in two variables x_1^2

or $(x_1-x_2)^2$ are positive semidefinite while $-x_1^2-x_2^2$ is negative definite and $x_1^2-x_2^2$ is indefinite.

Geometrically, Q positive definite means that the quadric surface

$$Q(x_1, x_2, \dots, x_n)=1$$

has an ellipsoidal character. Let us examine this in the two-dimensional case. Changing our notation we deal with the conic

$$(1) \qquad Ax^2+2Hxy+By^2=1.$$

If we make the change of variables

$$x=\xi\cos\theta-\eta\sin\theta$$

$$y=\xi\sin\theta+\eta\cos\theta$$

corresponding to a rotation of the (x, y) axes through an angle θ, we obtain a quadratic form in ξ, η

$$a\xi^2+2h\xi\eta+b\eta^2=1$$

where

$$2h=(B-A)\sin 2\theta+2H\cos 2\theta.$$

If we choose θ so that $\tan 2\theta=2H/(A-B)$ then we obtain

$$a\xi^2+b\eta^2=1.$$

This transformation can be expressed in matrix form as an orthogonal similarity:

$$\begin{bmatrix} c & s \\ -s & c \end{bmatrix}\begin{bmatrix} A & H \\ H & B \end{bmatrix}\begin{bmatrix} c & -s \\ s & c \end{bmatrix}=\begin{bmatrix} a & 0 \\ 0 & b \end{bmatrix}.$$

Since

$$Ax^2+2Hxy+By^2=A^{-1}(Ax+Hy)^2+A^{-1}(AB-H^2)y^2$$

this form is positive definite (in the sense of Def. 1) if and only if $A>0$ and $AB-H^2>0$ (and this implies $B>0$). In this case the characteristic values a, b of the matrix $\begin{bmatrix} A & H \\ H & B \end{bmatrix}$ are positive and the conic (1), when referred to the rotated axes, has equation

$$\frac{\xi^2}{\alpha^2}+\frac{\eta^2}{\beta^2}=1, \quad \alpha^2=a^{-1}, \quad \beta^2=b^{-1}.$$

Our object in this section is to justify two definitions equivalent to Def. 1 above.

Def. 2. *A symmetric matrix* **A** *is positive definite if all its characteristic values are positive.*

Def. 3. *A symmetric matrix* $\boldsymbol{A}$ *is positive definite if all its leading principal minors are positive.*

We shall use the following fact concerning a symmetric matrix $\boldsymbol{A}$: We can find a (real) orthogonal matrix $\boldsymbol{O}$ such that

$$\boldsymbol{O}'\boldsymbol{A}\boldsymbol{O} = \operatorname{diag}[\alpha_1, \alpha_2, \ldots, \alpha_n]. \tag{2}$$

Since $\boldsymbol{O}'\boldsymbol{O} = \boldsymbol{I}$ we have $\boldsymbol{O}' = \boldsymbol{O}^{-1}$ and (2) is a similarity so that $\alpha_1, \alpha_2, \ldots, \alpha_n$ are the characteristic values of $\boldsymbol{A}$ and are real.

1⇒2.

If α is a characteristic value of $\boldsymbol{A}$ and $\boldsymbol{a}$ the corresponding characteristic vector we have

$$\boldsymbol{A}\boldsymbol{a} = \alpha\boldsymbol{a}$$

and hence

$$\boldsymbol{a}'\boldsymbol{A}\boldsymbol{a} = \alpha\boldsymbol{a}'\boldsymbol{a}$$

so that

$$\alpha = \boldsymbol{a}'\boldsymbol{A}\boldsymbol{a}/\boldsymbol{a}'\boldsymbol{a}$$

which is necessarily positive since $\boldsymbol{a} \neq 0$, $\boldsymbol{A}$ being positive definite in the sense of Def. 1.

2⇒1.

We now assume that each characteristic value α_i of $\boldsymbol{A}$ is positive and use (2) to get

$$\begin{aligned} \boldsymbol{x}'\boldsymbol{A}\boldsymbol{x} &= \boldsymbol{x}'\boldsymbol{O}\boldsymbol{O}'\boldsymbol{A}\boldsymbol{O}\boldsymbol{O}'\boldsymbol{x} \\ &= \boldsymbol{y}' \operatorname{diag}[\alpha_1, \alpha_2, \ldots, \alpha_n]\boldsymbol{y} \\ &= \sum \alpha_i y_i^2 \end{aligned}$$

where $\boldsymbol{y} = \boldsymbol{O}'\boldsymbol{x}$. Since each $\alpha_i > 0$ it is clear that $\boldsymbol{x}'\boldsymbol{A}\boldsymbol{x} \geqq 0$ and equality can only occur if $\boldsymbol{y} = 0$ and this means $\boldsymbol{x} = 0$. Hence $\boldsymbol{A}$ is positive definite in the sense of Def. 1.

1⇒3.

In view of the equivalence of 1 and 2 and of the fact that $\det \boldsymbol{A} = \pi\alpha_i$ we conclude that the determinant of a positive definite matrix is necessarily positive. We note that *any* principal submatrix $\hat{\boldsymbol{A}}$ of a positive definite matrix is positive definite. If $\hat{\boldsymbol{A}}$ is $k \times k$ then for any k-dimensional vector $\hat{\boldsymbol{x}}$ we have

$$\hat{\boldsymbol{x}}'\hat{\boldsymbol{A}}\hat{\boldsymbol{x}} = \boldsymbol{x}'\boldsymbol{A}\boldsymbol{x}$$

where $\boldsymbol{x}$ is the n-dimensional vector formed by "completing" $\hat{\boldsymbol{x}}$ with zeros.

Hence $\mathbf{x}'\hat{\mathbf{A}}\hat{\mathbf{x}} \geqq 0$ and there is equality only if $\hat{\mathbf{x}}=0$. It follows, as before, that $\det \hat{\mathbf{A}} > 0$, so that $\mathbf{A}$ is certainly positive definite in the sense of Def. 3.

3⇒1.

We now assume only that the *leading* principal minors of $\mathbf{A}$ are positive. It follows from the LDU theorem that we can factorize $\mathbf{A}$ as $\mathbf{A}=\mathbf{L}\mathbf{L}'$. We begin by showing that L is *real.* This is done by repeating the (induction) proof of the LDU theorem, taking account of additional hypotheses.

For $n=1$ the result is trivial since $a_{11}>0$, and we have the triangular decomposition: $\mathbf{A}=[a_{11}]=[\sqrt{a_{11}}]\,[\sqrt{a_{11}}]$.

In the general step we assume a real factorization $\hat{\mathbf{A}}=\hat{\mathbf{L}}\hat{\mathbf{L}}'$ and we consider the factorization

$$\mathbf{A}=\begin{bmatrix}\hat{\mathbf{A}} & \hat{\boldsymbol{a}}' \\ \hat{\boldsymbol{a}} & a_{nn}\end{bmatrix}=\begin{bmatrix}\hat{\mathbf{L}} & 0 \\ \hat{\mathbf{x}} & l_{nn}\end{bmatrix}\begin{bmatrix}\hat{\mathbf{L}}' & \hat{\mathbf{x}}' \\ 0 & l_{nn}\end{bmatrix}. \tag{3}$$

Since $\hat{\mathbf{A}}$ is non-singular ($\det \hat{\mathbf{A}}>0$) it follows that $\hat{\mathbf{L}}$ is non-singular and hence we find that $\hat{\mathbf{x}}=\hat{\boldsymbol{a}}\hat{\mathbf{L}}'^{-1}$ is real. Now, taking determinants,

$$\det \mathbf{A}=[l_{nn}\det \hat{\mathbf{L}}]^2$$

and so, $\det \mathbf{A}$ being positive, we conclude that l_{nn} is real and the factorization (3) is real.

We complete the proof 3⇒1 as follows. We have

$$\mathbf{x}'\mathbf{A}\mathbf{x}=\mathbf{x}'\mathbf{L}\mathbf{L}'\mathbf{x}=\mathbf{y}'\mathbf{y}$$

where $\mathbf{y}=\mathbf{L}'\mathbf{x}$ is real. Hence $\mathbf{x}'\mathbf{A}\mathbf{x}\geqq 0$ and we can have equality only if $\mathbf{y}=0$ which means $\mathbf{x}=0$, $\mathbf{L}$ being non-singular.

Remarks

(1) If we only assume $\mathbf{A}$ non-negative definite, representations of $\mathbf{A}$ as $\mathbf{A}=\mathbf{L}\mathbf{L}'$ need not be unique, e.g.,

$$\begin{bmatrix}0 & 0 \\ 0 & 1\end{bmatrix}=\begin{bmatrix}0 & 0 \\ c & s\end{bmatrix}\begin{bmatrix}0 & c \\ 0 & s\end{bmatrix}$$

where $c=\cos\theta$, $s=\sin\theta$, for any θ.

(2) We note that a direct application of the $\mathbf{L}\mathbf{L}'$ factorization will produce representation of $\mathbf{Q}=\mathbf{x}'\mathbf{A}\mathbf{x}$ as a "sum" of squares. Assuming that the process is feasible we find

$$\mathbf{A}=\mathbf{L}\mathbf{D}\mathbf{L}'$$

with $\mathbf{L}$, $\mathbf{D}$ real and $\mathbf{L}$ has a unit diagonal. Hence

$$\begin{aligned} \boldsymbol{x}'\mathbf{A}\boldsymbol{x} &= \boldsymbol{x}'\mathbf{L}\mathbf{D}\mathbf{L}'\boldsymbol{x} \\ &= \boldsymbol{y}'\mathbf{D}\boldsymbol{y} \\ &= \sum d_{ii} y_i^2 \end{aligned}$$

where

$$\boldsymbol{y} = \mathbf{L}'\boldsymbol{x}, \quad \text{i.e.,} \quad y_i = x_i + \sum_{j>i} l_{ji} x_j, \quad i = 1, 2, \ldots, n.$$

This transformation is obtained by linear computations and, in general, will not be an orthogonal one. We have already remarked that an orthogonal reduction of the quadratic form to a sum of squares, i.e., a diagonalization of the matrix is always possible in the symmetric case but this involves getting the characteristic values, in principle solving an equation of degree n.

Chapter 6, Problems

6.1. Show that a left characteristic vector, $\boldsymbol{r}$, for $\mathbf{A}$ corresponding to a characteristic value α is orthogonal to a right characteristic vector $\boldsymbol{c}$ of $\mathbf{A}$ which corresponds to a characteristic value $\beta \neq \alpha$.

6.2 What are the characteristic values of the anti-scalar matrix $\boldsymbol{x}\boldsymbol{y}'$? What is its rank?

Let $k_1, k_2, \ldots, k_n$ be any non-zero numbers. Let

$$\mathbf{A} = [a_{ij}] \quad \text{where} \quad a_{ij} = k_i/k_j, \quad i, j = 1, 2, \ldots, n.$$

What are the characteristic values of $\mathbf{A}$?

Find an orthogonal matrix which diagonalises $\boldsymbol{x}\boldsymbol{x}'$. Can you find an orthogonal matrix which diagonalises $\boldsymbol{x}\boldsymbol{y}'$?

6.3. Show by examples that if α_i, β_i are the characteristic values of matrices $\mathbf{A}$ and $\mathbf{B}$ it is not necessarily true that the characteristic values of $\mathbf{AB}$ and $\mathbf{A}+\mathbf{B}$ are $\alpha_i\beta_i$ and $\alpha_i+\beta_i$, no matter how ordered.

Give conditions on $\mathbf{A}$ and $\mathbf{B}$ which are sufficient to ensure that this is the case.

6.4. If α is a characteristic value of $\mathbf{A}$ and $\boldsymbol{a}$ a corresponding characteristic vector, show that $\boldsymbol{a}$ is a characteristic vector of $p(\mathbf{A})$, where p is any polynomial and find the corresponding characteristic value. Suppose $\mathbf{A}$ is non-singular and that α, $\boldsymbol{a}$ is a characteristic pair for $\mathbf{A}$. Show that α^{-1}, $\boldsymbol{a}$ is a characteristic pair for $\mathbf{A}^{-1}$.

State a result about $r(\mathbf{A})$, where $r(x)$ is a rational function of x (i.e., the quotient of two polynomials).

6.5. Let $\mathbf{P}=\begin{bmatrix}0&1&0&0\\0&0&1&0\\0&0&0&1\\1&0&0&0\end{bmatrix}$. Compute $\mathbf{P}^2$, $\mathbf{P}^3$, $\mathbf{P}^4$ and $\mathbf{Q}=a\mathbf{I}+b\mathbf{P}+c\mathbf{P}^2+d\mathbf{P}^3$.

What is the characteristic polynomial of $\mathbf{P}$ and what are its characteristic values and vectors? What are the characteristic values of $\mathbf{Q}$?

State corresponding results for the circulant matrix of order $(n+1)$

$$\begin{bmatrix}a_0 & a_1 & a_1 \dots a_n\\ a_n & a_0 & a_2 \dots a_{n-1}\\ & & \dots \\ a_1 & a_2 & a_3 \dots a_0\end{bmatrix}.$$

Discuss in particular, the matrix $\mathbf{R}=\begin{bmatrix}-2&1&0&1\\1&-2&1&0\\0&1&-2&1\\1&0&1&-2\end{bmatrix}$.

6.6. Let $\mathbf{A}$, $\mathbf{B}$ be two $n\times n$ matrices. Show that the characteristic values of the $2n\times 2n$ matrix

$$\mathcal{A}=\begin{bmatrix}\mathbf{A}&\mathbf{B}\\\mathbf{B}&\mathbf{A}\end{bmatrix}$$

are those of the $n\times n$ matrices $\mathbf{A}+\mathbf{B}$ and $\mathbf{A}-\mathbf{B}$.

Find the characteristic vectors of $\mathcal{A}$ in terms of those of $\mathbf{A}+\mathbf{B}$ and $\mathbf{A}-\mathbf{B}$.

6.7. Draw the Gerschgorin circles for the matrices

$$\mathbf{A}=\begin{bmatrix}0&2\\-1&3\end{bmatrix},\quad \mathbf{B}=\begin{bmatrix}0&1\\1&3\end{bmatrix},$$

and mark on the diagrams the position of their characteristic values.

If $\mathbf{X}=\operatorname{diag}\,[1,\tfrac{1}{2}]$ compute $\mathbf{A}_1=\mathbf{X}^{-1}\mathbf{A}\mathbf{X}$ and draw the Gerschgorin circles for $\mathbf{A}_1$. If $\mathbf{Y}=\operatorname{diag}\,[1,(3-\sqrt{5})/2]$ draw the Gerschgorin circles for $\mathbf{Y}^{-1}\mathbf{B}\mathbf{Y}$.

6.8. (P. Furtwängler.) Establish the dominant diagonal theorem by induction.

6.9. If $\boldsymbol{A}$ is a positive definite matrix show that $\varrho(\boldsymbol{A}) \geqq \max a_{ii}$. Using the Perron—Frobenius theorem show that when $\boldsymbol{B}$ is a positive matrix, then

$$\varrho(\boldsymbol{B}) > \max b_{ii},$$

unless when $\boldsymbol{B}$ is a 1×1 matrix.

6.10. Show that the matrix

$$\begin{bmatrix} 28 & 6 & 6\sqrt{3} \\ 6 & 13 & 5\sqrt{3} \\ 6\sqrt{3} & 6\sqrt{3} & 23 \end{bmatrix}$$

has characteristic roots 5, 2, 1. Find the corresponding characteristic vectors; verify that they are orthogonal.

6.11. Suppose $\boldsymbol{A}$ is a non-singular symmetric matrix with distinct characteristic values. The "modal matrix" $\boldsymbol{M}$ of $\boldsymbol{A}$ is the matrix formed by the characteristic vectors of $\boldsymbol{A}$. Show that $\boldsymbol{M}\boldsymbol{M}'$ is diagonal (and, by suitable normalization, $\boldsymbol{M}\boldsymbol{M}' = \boldsymbol{I}$ so that $\boldsymbol{M}$ is orthogonal).

What are corresponding results when we do not assume symmetry?

6.12. Show that if $\boldsymbol{A}$, $\boldsymbol{B}$ are $n \times n$ matrices of which one at least is non-singular then $\boldsymbol{A}\boldsymbol{B}$ and $\boldsymbol{B}\boldsymbol{A}$ are similar and therefore have the same characteristic values. (This result is true without the non-singularity condition.) Formulate a corresponding result when $\boldsymbol{A}$, $\boldsymbol{B}$ are rectangular matrices.

If $\boldsymbol{A}$ is a 3×2 matrix and $\boldsymbol{B}$ is a 2×3 matrix and

$$\boldsymbol{A}\boldsymbol{B} = \begin{bmatrix} 8 & 2 & -2 \\ 2 & 5 & 4 \\ -2 & 4 & 5 \end{bmatrix}$$

show that

$$\boldsymbol{B}\boldsymbol{A} = \begin{bmatrix} 9 & 0 \\ 0 & 9 \end{bmatrix}.$$

6.13. Is the matrix $\boldsymbol{A} = \begin{bmatrix} 0 & 1 & 1 \\ 1 & 0 & 1 \\ 1 & 1 & 0 \end{bmatrix}$ positive definite or not?

Express the quadratic form

$$Q = \boldsymbol{x}'\boldsymbol{A}\boldsymbol{x} = 2x_1x_2 + 2x_2x_3 + 2x_3x_1$$

as a "sum" of squares.

6.14. The generalized characteristic value problem (g.c.v.p.) is the solution of

$$\mathbf{A}\mathbf{x}=\lambda\mathbf{B}\mathbf{x}$$

for λ, $\mathbf{x}$ when $\mathbf{A}$, $\mathbf{B}$ are given. If $\mathbf{B}=\mathbf{I}$ it is the ordinary characteristic value problem for $\mathbf{A}$, and if $\mathbf{B}$ is non-singular, it reduces to the ordinary characteristic value problem for $\mathbf{B}^{-1}\mathbf{A}$.

Solve the g.c.v.p. completely in the case when $\mathbf{A}$, $\mathbf{B}$ are the following triple diagonal matrices

$$\mathbf{A}=[\ldots, \overset{\cdots}{\underset{\cdots}{a, b, a}}, \ldots], \quad \mathbf{B}=[\ldots, \overset{\cdots}{\underset{\cdots}{\alpha, \beta, \alpha}}, \ldots]$$

giving the general form for λ and for $\mathbf{x}$. Are any restrictions on a, b, α, β required?

Write out completely the values of λ and $\mathbf{x}$ in the case when $\mathbf{A}$, $\mathbf{B}$ are 4×4 matrices and

$$a=-5, \quad b=10 \quad \text{and} \quad \beta=2/15, \quad \alpha=1/30.$$

(Use Problem 6.4.)

6.15. What are the characteristic values of the (complex) matrix

$$\begin{bmatrix} 0 & \mathbf{A}^* \\ \mathbf{A} & 0 \end{bmatrix},$$

where $\mathbf{A}^*=\bar{\mathbf{A}}'$?

6.16. (Williamson.) Let $\mathcal{A}$ be an $mn\times mn$ matrix partitioned into m^2 sub-matrices $\mathbf{A}_{ij}$, $i, j=1, 2, \ldots, m$ each $n\times n$. Suppose that each $\mathbf{A}_{ij}$ is a polynomial $a_{ij}(\mathbf{A})$ in a fixed $n\times n$ matrix $\mathbf{A}$. Then the mn characteristic values of $\mathcal{A}$ can be obtained as follows: for each characteristic value α_k, $k=1, 2, \ldots, n$, of $\mathbf{A}$ form the $m\times m$ matrix

$$[a_{ij}(\alpha_k)]$$

and find its m characteristic values α_{kl}, $l=1, 2, \ldots, m$. The α_{kl} are the required characteristic values of $\mathcal{A}$.

6.17. Let $\mathbf{c}$ be an n-dimensional column vector and $\mathbf{r}'$ an n-dimensional row vector. Let $\mathbf{A}=[a_{ij}]$ be defined by

$$a_{ij}=c_i c_j+r_i r_j.$$

What is the rank of $\mathbf{A}$ and what is its characteristic polynomial?

6.18. Show that the matrix

$$A=\frac{1}{3}\begin{bmatrix} 1 & -2 & 2 \\ -2 & 1 & 2 \\ -2 & -2 & 1 \end{bmatrix}$$

is orthogonal. Find its characteristic pairs.

Is A orthogonally similar to a diagonal matrix? Is A unitarily similar to a diagonal matrix?

6.19. (Kantorovich.) Let A be a positive definite hermitian matrix with characteristic values $\lambda_1 \geqq \lambda_2 \geqq \ldots \geqq \lambda_n > 0$. Show that for every vector $x \neq 0$

$$1 = \frac{(x^* A x)(x^* A^{-1} x)}{(x^* x)^2} \leqq \frac{(\varkappa^{1/2}(A) + \varkappa^{-1/2}(A))^2}{4},$$

where $\varkappa(A)$ is the euclidean condition number of A.

CHAPTER 7

The Power Method, Deflation, Inverse Iteration

We now discuss a method for determining the dominant characteristic root of a matrix, it being assumed that there is only one α_i, say α_1, with $|\alpha_i|=\varrho(\boldsymbol{A})$. This method, then, will always work when $\boldsymbol{A}>0$.

We then show, once we have found a characteristic value and vector of a matrix $\boldsymbol{A}$, how to find a "deflated" $(n-1)\times(n-1)$ matrix $\boldsymbol{A}_1$ which has for its characteristic roots the *other* characteristic roots of $\boldsymbol{A}$ and for which there is a simple relation between the characteristic vectors of $\boldsymbol{A}$, $\boldsymbol{A}_1$.

We conclude this chapter by showing how to find the characteristic vector of a matrix corresponding to a given characteristic value by the method of inverse iteration. This is particularly convenient when the matrix is a triple diagonal one — in Chapter 8 we shall show how to reduce the characteristic value problem for a symmetric matrix to the triple diagonal case.

1. The Power Method

For simplicity assume that the characteristic vectors $\boldsymbol{c}_i$, $i=1,2,\dots,n$, span the whole space $\boldsymbol{R}_n$. This is the case, e.g., when $\boldsymbol{A}$ is symmetric. Take any vector $\boldsymbol{v}^{(0)}$ and represent it in terms of the basis $\{\boldsymbol{c}_i\}$:

$$\boldsymbol{v}^{(0)}=\sum a_i \boldsymbol{c}_i .$$

We shall assume that $\boldsymbol{v}^{(0)}$ is normalized, e.g., to have $\|\boldsymbol{v}^{(0)}\|_2=1$. Then we form

$$\begin{aligned}\boldsymbol{A}\,\boldsymbol{v}^{(0)}&=\sum a_i \boldsymbol{A}\,\boldsymbol{c}_i=\sum a_i \alpha_i \boldsymbol{c}_i\\ &=\mu^{(1)}\,\boldsymbol{v}^{(1)}, \quad \text{say,}\end{aligned}$$

where $\|\boldsymbol{v}^{(1)}\|_2=1$. We proceed to get

$$\boldsymbol{A}\,\boldsymbol{v}^{(1)}=\mu^{(2)}\,\boldsymbol{v}^{(2)}$$

where $\|\boldsymbol{v}^{(2)}\|_2=1$ and so on. We assert that, provided $a_1\neq 0$, we have

$$\mu^{(r)}\to\alpha_1=\varrho(\boldsymbol{A}) \quad \text{and} \quad \boldsymbol{v}^{(r)}\to\boldsymbol{c}_1 .$$

It is clear that $\boldsymbol{v}^{(r)}$ is obtainable by normalizing

$$\begin{aligned}\boldsymbol{A}^r\,\boldsymbol{v}^{(0)}&=\sum a_i \alpha_i^r \boldsymbol{c}_i\\ &\doteqdot a_1 \alpha_1^r \boldsymbol{c}_1\end{aligned}$$

if $a_1\neq 0$. Hence $\boldsymbol{v}^{(r)}\to\boldsymbol{c}_1$ and $\mu^{(r)}\to\alpha_1$.

The condition that $a_1 \neq 0$ is not critical; if $a_1 = 0$ then during the calculation a component of c_1 may be introduced through round-off and it will be amplified and will ultimately predominate.

The behavior in a good case, i.e., when the dominant α is well separated from the others, is shown in the following example.

$$\mathbf{A} = \begin{bmatrix} .2 & .9 & 1.32 \\ -11.2 & 22.28 & -10.72 \\ -5.8 & 9.45 & -1.94 \end{bmatrix}, \quad \boldsymbol{v}^{(0)} = \begin{bmatrix} 1 \\ 0 \\ 0 \end{bmatrix}.$$

We shall normalize the vectors by making these first coefficient unity instead of making their length unity.

We obtain the following results:

$$\mathbf{A}\boldsymbol{v}^{(0)} = [.2, \ -11.2, \ -5.8]' = \mu^{(1)}\boldsymbol{v}^{(1)} \quad \text{where} \quad \mu^{(1)} = .2, \quad \boldsymbol{v}^{(1)} = [1, \ -56, \ -29]'$$

$$\mathbf{A}\boldsymbol{v}^{(1)} = [-88.48, \ -948, \ -478.7]' = \mu^{(2)}\boldsymbol{v}^{(2)} \quad \text{where} \quad \mu^{(2)} = -88.48,$$

$$\boldsymbol{v}^{(2)} = [1, \ 10.7143, \ 5.4107]'.$$

2. Deflation

There are several systematic ways of deflating an $n \times n$ matrix. Any one of these, in principle, would enable us to determine by repeated application all the characteristic values and vectors of $\mathbf{A}$, provided these are separated: $|\alpha_1| > |\alpha_2| > \ldots > |\alpha_n|$. However this is not a very practical one for the following reasons. First we will not get α_1 exactly, and so the machine $\mathbf{A}_1$ will be different from the true $\mathbf{A}_1$: then the α_2 determined will be different from the true dominant characteristic value of the machine $\mathbf{A}_1$. Thus we get a successive contamination of our answers and the method is practical only for the first few α's.

We discuss one method of deflation; another is suggested in Problem 7.8.

Suppose $\boldsymbol{x} = [1, \hat{\boldsymbol{x}}]'$ is a characteristic vector of $\mathbf{A}$ corresponding to the characteristic value α so that

$$\begin{bmatrix} a_{11} & \hat{\boldsymbol{r}}' \\ \hat{\boldsymbol{c}} & \hat{\mathbf{A}} \end{bmatrix} \begin{bmatrix} 1 \\ \hat{\boldsymbol{x}} \end{bmatrix} = \alpha \begin{bmatrix} 1 \\ \hat{\boldsymbol{x}} \end{bmatrix} \quad \text{which gives} \quad \left.\begin{aligned} a_{11} + \hat{\boldsymbol{r}}'\hat{\boldsymbol{x}} &= \alpha \\ \hat{\boldsymbol{c}} + \hat{\mathbf{A}}\hat{\boldsymbol{x}} &= \alpha\hat{\boldsymbol{x}} \end{aligned}\right\}.$$

(We use ^ systematically to indicate vectors of dimension $n-1$ or matrices of order $n-1$.)

If

$$\mathbf{S} = \begin{bmatrix} 1 & \hat{0}' \\ \hat{\boldsymbol{x}} & \mathbf{I} \end{bmatrix}$$

we find that

$$\mathbf{S}^{-1}=\begin{bmatrix} 1 & \hat{0}' \\ -\hat{x} & \mathbf{I} \end{bmatrix}$$

and

$$\mathbf{S}^{-1}\mathbf{A}\mathbf{S}=\begin{bmatrix} a_{11}+\hat{r}'\hat{x} & \hat{r}' \\ -a_{11}\hat{x}+\hat{c}-\hat{x}\hat{r}'\hat{x}+\hat{\mathbf{A}}\hat{x} & \hat{\mathbf{A}}-\hat{x}\hat{r}' \end{bmatrix}$$

which simplifies to an upper triangular matrix

$$\mathbf{S}^{-1}\mathbf{A}\mathbf{S}=\begin{bmatrix} \alpha & \hat{r}' \\ 0 & \hat{\mathbf{A}}-\hat{x}\hat{r}' \end{bmatrix}.$$

Write $\mathbf{A}_1=\hat{\mathbf{A}}-\hat{x}\hat{r}'$.

Suppose $\mathbf{S}^{-1}\mathbf{A}\mathbf{S}$ has a characteristic vector $y=[1, \hat{y}]'$ corresponding to a characteristic value $\lambda\neq\alpha$ so that

$$(1) \qquad \mathbf{S}^{-1}\mathbf{A}\mathbf{S}\begin{bmatrix} 1 \\ \hat{y} \end{bmatrix}=\lambda\begin{bmatrix} 1 \\ \hat{y} \end{bmatrix} \quad \text{which gives} \quad \left.\begin{aligned} \alpha+\hat{r}'\hat{y}&=\lambda \\ \mathbf{A}_1\hat{y}&=\lambda\hat{y} \end{aligned}\right\}.$$

Let $\tilde{y}, \lambda$ be *any* characteristic pair for $\mathbf{A}_1$.

(a) Then if $\hat{r}'\tilde{y}\neq 0$ we can verify

$$\hat{y}=\frac{\lambda-\alpha}{\hat{r}'\tilde{y}}\cdot\tilde{y}$$

satisfies (1) and $[1, \hat{y}']'$, λ is a characteristic pair for $\mathbf{S}^{-1}\mathbf{A}\mathbf{S}$ and

$$\mathbf{S}y=\begin{bmatrix} 1 & \hat{0}' \\ \hat{x} & \mathbf{I} \end{bmatrix}\begin{bmatrix} 1 \\ (\lambda-\alpha)\tilde{y}/(\hat{r}'\tilde{y}) \end{bmatrix}=\begin{bmatrix} 1 \\ \hat{x}+\{(\lambda-\alpha)\tilde{y}/(\hat{r}'\tilde{y})\} \end{bmatrix}$$

is the characteristic vector of $\mathbf{A}$ corresponding to λ.

(b) If $\hat{r}'\tilde{y}=0$ then $[0, \tilde{y}']'$ is a characteristic vector of $\mathbf{S}^{-1}\mathbf{A}\mathbf{S}$ and also of $\mathbf{A}$, corresponding to λ.

Thus we have a deflation process which enables us to get from a characteristic pair x, α of a $n\times n$ matrix $\mathbf{A}$ to an $(n-1)\times(n-1)$ matrix $\mathbf{A}_1$. If we can find a characteristic pair $\tilde{y}$, λ for $\mathbf{A}_1$ with $\lambda\neq\alpha$ then we can get a new characteristic pair for $\mathbf{A}$.

See Problem 7.7 for an example.

3. Wielandt inverse iteration

In general, if we know a characteristic value α of $\mathbf{A}$ so that $(\mathbf{A}-\alpha\mathbf{I})x=0$ is a homogeneous system of equations with a non-trivial solution we find x by discarding one of the equations and then solving. It is not difficult to construct (even triple diagonal) examples where the results are most unsatisfactory.

The following process is recommended: Take any vector $\boldsymbol{v}^{(0)}$ and solve for $\boldsymbol{v}^{(1)}, \boldsymbol{v}^{(2)}, \ldots$ from the equations

$$(2) \qquad (\boldsymbol{A}-\tilde{\alpha}\boldsymbol{I})\boldsymbol{v}^{(r+1)}=\boldsymbol{v}^{(r)},$$

where $\tilde{\alpha}$ is our approximation to the characteristic value α_1 under consideration: it will be found that the vectors $\boldsymbol{v}^{(r)}$ will very quickly approach the characteristic vector $\boldsymbol{c}_1$. The reasoning for this is similar to that used to justify the power method, but note that we need not assume α_1 to be dominant. If

$$\boldsymbol{v}^{(0)}=\sum_{i=1}^{n} a_i \boldsymbol{c}_i$$

then

$$(3) \qquad \boldsymbol{v}^{(r+1)}=\frac{1}{(\alpha_1-\tilde{\alpha})^r} a_1 \boldsymbol{c}_1+\ldots$$

and if $\tilde{\alpha}$ is closer to α_1 than to any of the other α's the first term in (3) will predominate and the $\boldsymbol{v}^{(r)}$ will approach $\boldsymbol{c}_1$, the faster the nearer the approximate $\tilde{\alpha}$ is to α_1.

If $\boldsymbol{A}$ is a triple diagonal matrix, so is $\boldsymbol{A}-\hat{\alpha}\boldsymbol{I}$, and the solution of the equations for the $\boldsymbol{v}^{(r)}$ can be accomplished quickly.

Chapter 7, Problems

7.1. Continue the example in the text and compute

$$\mu^{(3)}= \quad , \boldsymbol{v}^{(3)}=[1, \qquad , \qquad]'$$

and

$$\mu^{(4)}= \quad , \boldsymbol{v}^{(4)}=[1, \qquad , \qquad]'$$

by desk calculator; then write a program for doing this and use it to find $\mu^{(10)}$, $\boldsymbol{v}^{(10)}$. Then find α_1 and $\boldsymbol{v}_1$ exactly.

7.2. Find the dominant characteristic value and vector for

$$\begin{bmatrix} 133 & 6 & 135 \\ 44 & 5 & 46 \\ -88 & -6 & -90 \end{bmatrix}$$

by the power method.

Find also the other characteristic values and vectors.

7.3. Use the program just prepared on the following bad example due to Bodewig:

$$\boldsymbol{A}=\begin{bmatrix}2 & 1 & 3 & 4\\ 1 & -3 & 1 & 5\\ 3 & 1 & 6 & -2\\ 4 & 5 & -2 & -1\end{bmatrix}, \qquad \boldsymbol{v}^{(0)}=\begin{bmatrix}1\\1\\1\\1\end{bmatrix}.$$

(Arrange to print out $\mu^{(r)}$, $\boldsymbol{v}^{(r)}$ for $r=0$ (50) 500, e.g.)

7.4. Apply the power method to find the dominant characteristic vector of $\boldsymbol{H}_4$, starting with $\boldsymbol{v}^{(0)}=[1, .6, .4, .3]'$. Use the Rayleigh quotient to get the dominant characteristic value.

7.5. Discuss the power method in the case of a 2×2 matrix $\boldsymbol{A}$.

7.6. A 4×4 matrix is known to have characteristic values near 20, 10, 5, 1. The power method will therefore enable us to find the root near 20. Show that the same method, when applied to a suitably shifted matrix $\boldsymbol{A}-\alpha\boldsymbol{I}$, will enable the characteristic value near 1 to be determined.

Apply this idea to find the smallest characteristic value of $\boldsymbol{H}_4$ and of the matrices of Problems 7.1, 7.2, shows.

7.7. Carry out in detail one step of the deflation process described on p. 66 starting off with the characteristic pair $[1, 0, 1]'$, 1 of the matrix

$$\begin{bmatrix}5 & 1 & -4\\ -2 & 2 & 2\\ 1 & 1 & 0\end{bmatrix}$$

and using the characteristic pair $[1, 1]'$, 4 of the deflated matrix.

7.8. Suppose $\boldsymbol{v}=[v_1, v_2, 1]'$ is a characteristic vector of a matrix $\boldsymbol{A}$ corresponding to a characteristic value λ_3. Prove that the other characteristic values of $\boldsymbol{A}$ are those of the matrix:

$$\begin{bmatrix}a_{11}-v_1 a_{31}, & a_{12}-v_1 a_{32}\\ a_{21}-v_2 a_{31}, & a_{22}-v_2 a_{32}\end{bmatrix}$$

and use this to handle Problem 7.2 or a similar problem. How can you find the other characteristic vectors?

7.9. Illustrate the Wielandt inverse iteration by discussing in detail the matrix

$$\boldsymbol{C}_4=\begin{bmatrix}2 & -1 & 0 & 0\\ -1 & 2 & -1 & 0\\ 0 & -1 & 2 & -1\\ 0 & 0 & -1 & 2\end{bmatrix}.$$

Discuss the approximation of the characteristic vector corresponding to the characteristic value

$$\alpha_1 = 2 - 2\cos(\pi/5) \doteqdot 0.3820.$$

Choose $\boldsymbol{v}_0 = [1, 1, 1, 1]'$ and $\tilde{\alpha} = 0.4$.

7.10. Illustrate the Wielandt inverse iteration by discussing the matrix

$$\begin{bmatrix} 5 & 7 & 6 & 5 \\ 7 & 10 & 8 & 7 \\ 6 & 8 & 10 & 9 \\ 5 & 7 & 9 & 10 \end{bmatrix}.$$

Find the characteristic vectors corresponding to the characteristic values which are approximately

$$30, \quad 4, \quad 0.8, \quad 0.01.$$

From the characteristic vectors so found estimate the corresponding characteristic value using the Rayleigh quotient.

7.11. Discuss the application of the power method to the matrix

$$\boldsymbol{A} = \begin{bmatrix} 0 & -1 & 1 \\ 1 & 0 & -1 \\ -1 & 1 & 0 \end{bmatrix}$$

with starting vector $\boldsymbol{v}^{(0)} = [1, 1, -2]'$.

CHAPTER 8

Characteristic Values

In Chapter 7 we have shown how to get the dominant characteristic value and vector of a matrix $\boldsymbol{A}$ and, indeed, how to get a *few* of the largest characteristic values and vectors by deflation, noting that the accuracy falls off rapidly. In this chapter we discuss two classes of methods by means of which we can get *all* the characteristic values of a matrix at substantially the same accuracy. The characteristic values being found, the corresponding vectors can be obtained by inverse iteration, as described in Chapter 7.

1. Rotation methods: Jacobi, Givens, Householder

We begin by recalling some elementary properties of conics already discussed in Chapter 6. We saw there that the "transformation to principal axes" could be expressed in matrix notation in the form

$$\begin{bmatrix} c & s \\ -s & c \end{bmatrix} \begin{bmatrix} A & H \\ H & B \end{bmatrix} \begin{bmatrix} c & -s \\ s & c \end{bmatrix} = \begin{bmatrix} a & 0 \\ 0 & b \end{bmatrix},$$

i.e., the diagonalization of a symmetric 2×2 matrix by an orthogonal similarity.

We consider a *real symmetric* matrix $\boldsymbol{A}$. We would like to determine an orthogonal matrix $\boldsymbol{S}$ such that

$$\boldsymbol{S}^{-1}\boldsymbol{A}\boldsymbol{S} = \operatorname{diag}[\alpha_1, \alpha_2, \dots, \alpha_n].$$

In principle this is possible. We have only to find the characteristic vectors $\boldsymbol{c}_i$ of $\boldsymbol{A}$ and then take $\boldsymbol{S} = [\boldsymbol{c}_1, \boldsymbol{c}_2, \dots, \boldsymbol{c}_n]$, but it is hardly conceivable that we can find the $\boldsymbol{c}_i$ without first knowing the α_i. What we do is to recall that in the context of numerical algebra we should ask only for *approximations* to the α's, and so we ask for matrices $\boldsymbol{S}$ which make $\boldsymbol{S}^{-1}\boldsymbol{A}\boldsymbol{S}$ almost diagonal, say having the sum of the squares of the off diagonal elements

$$J(\boldsymbol{A}) = \sum_{i \neq j} a_{ij}^2$$

small. There are several ways of handling this problem.

We require the following lemma (see Problem 2.19).

Lemma. *The Frobenius or Schur norm of a matrix* $\mathbf{A}$ *is invariant under orthogonal similarities.*

The plan of Jacobi is the following. Scan the off-diagonal a_{ij} and find one of largest absolute value. For simplicity suppose it is $a_{12} \neq 0$. Consider now the effect of an orthogonal similarity by

$$\mathbf{R} = \mathbf{R}_{12} = \left[\begin{array}{cc|c} c & -s & \\ s & c & 0 \\ \hline & 0 & \mathbf{I}_{n-2} \end{array}\right]$$

where c, s are chosen to make

$$\begin{bmatrix} c & s \\ -s & c \end{bmatrix} \begin{bmatrix} a_{11} & a_{12} \\ a_{21} & a_{22} \end{bmatrix} \begin{bmatrix} c & -s \\ s & c \end{bmatrix} = \begin{bmatrix} b_{11} & 0 \\ 0 & b_{22} \end{bmatrix}.$$

Using our lemma twice, first on 2×2 matrices and then on $n \times n$ matrices, we find:

$$a_{11}^2 + 2a_{12}^2 + a_{22}^2 = b_{11}^2 + b_{22}^2,$$

$$J(\mathbf{R}'\mathbf{A}\mathbf{R}) + b_{11}^2 + b_{22}^2 + a_{33}^2 + \ldots + a_{nn}^2 = J(\mathbf{A}) + a_{11}^2 + a_{22}^2 + \ldots + a_{nn}^2.$$

It follows that

$$J(\mathbf{R}'\mathbf{A}\mathbf{R}) = J(\mathbf{A}) - 2a_{12}^2.$$

By hypothesis

$$J(\mathbf{A}) \leqq (n^2 - n)\, a_{12}^2$$

so that

$$J(\mathbf{R}'\mathbf{A}\mathbf{R}) \leqq \left(1 - \frac{2}{n^2 - n}\right) J(\mathbf{A}).$$

We now recall the basic limit relation

$$\left(1 - \frac{x}{n}\right)^n \to e^{-x}.$$

It follows that we can reduce $J(\mathbf{A})$ by a factor k by repeating the above process, specifically by making about

$$\tfrac{1}{2} n^2 \log k^{-1}$$

rotations. Each rotation involves about $4n$ multiplications. This means that we can obtain reasonable estimates of all the characteristic values at an expense of $\mathcal{O}(n^3)$ operations. It is possible, in principle, to obtain the characteristic vectors at the same time: if we record the continued product of the rotations we approach the situation

$$\mathbf{S}^{-1}\mathbf{A}\mathbf{S}=\operatorname{diag}[\alpha_1, \ldots, \alpha_n]$$

and the columns of $\mathbf{S}$, the modal matrix of $\mathbf{A}$, are the characteristic vectors of $\mathbf{A}$.

There are certain disadvantages to this method. One is that after killing an off-diagonal element it does not remain dead — future rotations can restore this. Referring to the basic computation

$$\begin{bmatrix} \mathbf{O} & 0 \\ 0 & \mathbf{I} \end{bmatrix} \begin{bmatrix} \mathbf{A}_{11} & \mathbf{A}_{12} \\ \mathbf{A}_{21} & \mathbf{A}_{22} \end{bmatrix} \begin{bmatrix} \mathbf{O} & 0 \\ 0 & \mathbf{I} \end{bmatrix}$$

we see that only the elements in the 1, 2 rows and 1, 2 columns change. Similarly for a rotation which involves the r, s rows and r, s columns.

What can be done if we resolve never to touch an element which has been annihilated? Givens observed that we can reduce the general symmetric matrix to a *triple diagonal* one (and the general matrix to a so-called Hessenberg matrix) by annihilating in succession the elements in the

$$\begin{array}{r} (1, 3), (1, 4), \ldots, (1, n) \\ (2, 4), \ldots, (2, n) \\ \cdots \\ (n-2, n) \end{array}$$

positions by rotations involving the

$$\begin{array}{r} (2, 3), (2, 4), \ldots, (2, n) \\ (3, 4), \ldots, (3, n) \\ \cdots \\ (n-1, n) \end{array}$$

rows and columns. Thus the reduction to triple-diagonal form is done at the expense of

$$\tfrac{1}{2}(n-1)(n-2)$$

two-dimensional rotations and the characteristic value problem for the general symmetric matrix is reduced to that for a triple-diagonal matrix.

We have earlier noted that it is easy to compute the determinant of a triple diagonal matrix. That is, the value of the characteristic polynomial $p(x)$ for an assigned argument x is readily computed. From this observation several attacks on the characteristic value problem can be derived. We know that all the characteristic values are real and we can obtain bounds for these, e.g., via Gerschgorin's Theorem. Evaluation of $p(x)$ at the points obtained by continually bisecting this interval and, preferably, using Sturm's Theorem, enable the characteristic values to be obtained expeditiously.

The number of operations involved in the Givens reduction is easy to find. It is about

$$\tfrac{4}{3}n^3.$$

A significant improvement on this reduction process was given by Householder. Givens accomplishes the reduction by killing the elements one at a time by very simple orthogonal similarities: two-dimensional rotations. Householder does this by killing the elements, one *row* at a time by a more complicated orthogonal similarity. Detailed investigations show that *Householder's method takes about half as many operations* as Givens'. We require the following result. (See Problem 1.10.)

Lemma. *If ω is a vector of unit length so that $\omega'\omega=1$ then $\mathbf{I}-2\omega\omega'$ is orthogonal.*

Denote by ω_r an n-dimensional vector of unit length which has its first $(r-1)$ components zero:

$$\omega_r=[0,\ldots,0,x_r,x_{r+1},\ldots,x_n]',\; x_r^2+x_{r+1}^2+\ldots+x_n^2=1.$$

Write

$$\mathbf{P}_r=\mathbf{I}-2\omega_r\omega_r'.$$

We assert that ω_2 can be chosen so that

$$\mathbf{P}_2\mathbf{A}\mathbf{P}_2 \quad \text{has elements} \quad (1,3)(1,4),\ldots,(1,n)$$

(and their symmetric ones) zero; that $\mathbf{P}_3$ can be chosen so that

$$\mathbf{P}_3\mathbf{P}_2\mathbf{A}\mathbf{P}_2\mathbf{P}_3 \quad \text{has in addition the elements} \quad (2,4),\ldots,(2,n)$$

(and their symmetric ones) zero; etc.

It will be enough for our purposes to do this for the case of a (4×4) matrix

$$\mathbf{A}=\begin{bmatrix} a_1 & b_1 & c_1 & d_1\\ b_1 & b_2 & c_2 & d_2\\ c_1 & c_2 & c_3 & d_3\\ d_1 & d_2 & d_3 & d_4\end{bmatrix}.$$

In this case

$$\mathbf{P}_2=\mathbf{I}-2\begin{bmatrix} 0 & 0 & 0 & 0\\ 0 & x_2^2 & x_2x_3 & x_2x_4\\ 0 & x_2x_3 & x_3^2 & x_3x_4\\ 0 & x_2x_4 & x_3x_4 & x_4^2\end{bmatrix}.$$

Having regard to the structure of $\mathbf{P}_2$ it is clear that the (1, 3), (1, 4) elements of $\mathbf{P}_2\mathbf{A}\mathbf{P}_2$ are identical with the corresponding elements of $\mathbf{A}\mathbf{P}_2$. We show how to choose x_2, x_3, x_4 to make them zero.

We have

$$\mathbf{A}\mathbf{P}_2 = \mathbf{A} - 2\,\mathbf{A}\,\omega_2\,\omega_2'.$$

We have

$$\mathbf{A}\,\omega_2 = \begin{bmatrix} p_1 \\ p_2 \\ p_3 \\ p_4 \end{bmatrix} \quad \text{where} \quad p_1 = b_1 x_2 + c_1 x_3 + d_1 x_4.$$

The first row of $\mathbf{A}\mathbf{P}_2$ is:

$$[a_1,\ b_1 - 2p_1 x_2,\ c_1 - 2p_1 x_3,\ d_1 - 2p_1 x_4].$$

We want to have

$$\left.\begin{aligned} &(1) & c_1 - 2p_1 x_3 &= 0 \\ &(2) & d_1 - 2p_1 x_4 &= 0 \end{aligned}\right\}.$$

By orthogonality, the length of the first row of $\mathbf{A}$ is unchanged, so that

$$a_1^2 + b_1^2 + c_1^2 + d_1^2 = a_1^2 + (b_1 - 2p_1 x_2)^2 + 0 + 0$$

which we may write as

$$(3) \qquad b_1 - 2p_1 x_2 = \pm S^{1/2} \quad \text{where} \quad S = b_1^2 + c_1^2 + d_1^2.$$

If we multiply (1), (2) and (3) by x_3, x_4 and x_2 respectively and add we get

$$p_1 - 2p_1(x_2^2 + x_3^2 + x_4^2) = \pm x_2 S^{1/2}$$

so that

$$p_1 = \mp x_2 S^{1/2}.$$

The ambiguous sign is to be chosen as that of b_1, to make x_2 as large as possible, because of its occurrence in the denominators of x_3, x_4. All later ambiguities are the same as this first one.

Substituting back in (3) we get

$$b_1 \pm 2x_2^2 S^{1/2} = \pm S^{1/2}$$

and so

$$x_2^2 = \frac{1}{2}\left[1 \mp \frac{b_1}{S^{1/2}}\right].$$

Then we get

$$x_3 = \frac{c_1}{2p_1} = \frac{\mp c_1}{2x_2 S^{1/2}},$$

$$x_4 = \frac{\mp d_1}{2x_2 S^{1/2}}.$$

This establishes the basic step in the Householder reduction.

2. $\boldsymbol{LR}$ and $\boldsymbol{QR}$ Methods

We have seen that a matrix $\mathbf{A}_1$ which we shall assume to have all its leading submatrices non-singular can be represented as the form

$$\mathbf{A}_1=\mathbf{L}_1\mathbf{R}_1$$

where $\mathbf{L}_1$, $\mathbf{R}_1$ are non-singular lower and upper triangular matrices. We now compute $\mathbf{A}_2=\mathbf{R}_1\mathbf{L}_1$ and factorize it as

$$\mathbf{A}_2=\mathbf{L}_2\mathbf{R}_2$$

and continue in this way:

$$\mathbf{A}_n=\mathbf{L}_n\mathbf{R}_n,\quad \mathbf{A}_{n+1}=\mathbf{R}_n\mathbf{L}_n=\mathbf{L}_{n+1}\mathbf{R}_{n+1}.$$

In this section we shall assume $\mathbf{A}$ to be $m\times m$, so that n can be used as a current index.

We observe that

$$\mathbf{A}_{n+1}=\mathbf{R}_n(\mathbf{L}_n\mathbf{R}_n)\mathbf{R}_n^{-1}=\mathbf{R}_n\mathbf{A}_n\mathbf{R}_n^{-1}$$

so that all the $\mathbf{A}_n$ are similar. Rutishauser [1958] established that in certain circumstances the matrices $\mathbf{A}_n$ converge to an upper triangular matrix — and so, in principle, this is a way to get approximations to the characteristic values of $\mathbf{A}$. In practice this method is not too effective. Some of the drawbacks are the following: it is expensive for full matrices (see Problem 4.22); it is rather slowly convergent and it is often unstable numerically.

A variation on this $\mathbf{LR}$ algorithm, called the $\mathbf{QR}$ algorithm, was introduced by Francis and Kublanowskaja in 1961/62, and has been developed into a practically attractive procedure.

The main difference between the two algorithms is that in the second our factorizations are of the form:

$$\mathbf{A}_n=\mathbf{Q}_n\mathbf{R}_n,\quad \mathbf{A}_{n+1}=\mathbf{R}_n\mathbf{Q}_n=\mathbf{Q}_{n+1}\mathbf{R}_{n+1}, \tag{4}$$

where the $\mathbf{Q}_n$ are unitary and the $\mathbf{R}_n$ are upper triangular matrices. These factorizations are essentially the expression of the Gram-Schmidt orthonormalization process and so, in principle, are not different from the $\mathbf{LR}$ factorizations. The above representation is unique when we require that the diagonal elements of $\mathbf{R}$ to be positive. Cf. Problem 4.6.

It is not appropriate to give here a convergence proof nor to indicate the various devices which can accelerate convergence. We shall instead discuss the basic process in the case of a general unimodular 2×2 matrix.

From (4) we have

$$\mathbf{A}_{n+1}=\mathbf{R}_n\mathbf{Q}_n=(\mathbf{Q}_n^*\mathbf{Q}_n)\mathbf{R}_n\mathbf{Q}_n=\mathbf{Q}_n^*(\mathbf{Q}_n\mathbf{R}_n)\mathbf{Q}_n=\mathbf{Q}_n^*\mathbf{A}_n\mathbf{Q}_n$$

and, if we write $\mathbf{P}_n=\mathbf{Q}_1\mathbf{Q}_2\ldots\mathbf{Q}_n$ then $\mathbf{P}_n$ is unitary and

$$\mathbf{A}_{n+1}=\mathbf{P}_n^*\mathbf{A}\,\mathbf{P}_n. \tag{5}$$

If we write $\mathbf{S}_n=\mathbf{R}_n\mathbf{R}_{n-1}\ldots\mathbf{R}_1$, then $\mathbf{S}_n$ is upper triangular and it can be proved by induction that

$$\mathbf{P}_k\mathbf{S}_k=\mathbf{A}^k, \quad k=1,2,\ldots. \tag{6}$$

(For $k=1$, this is trivial. Assuming (6) we have, using (5),

$$\mathbf{P}_{k+1}\mathbf{S}_{k+1}=\mathbf{P}_k\mathbf{Q}_{k+1}\mathbf{R}_{k+1}\mathbf{S}_k=\mathbf{P}_k\mathbf{A}_{k+1}\mathbf{S}_k=\mathbf{P}_k(\mathbf{P}_k^*\mathbf{A}\,\mathbf{P}_k)\mathbf{S}_k=\mathbf{A}\cdot\mathbf{A}^k=\mathbf{A}^{k+1}.)$$

Thus it appears that if $\mathbf{P}_n$ is the unitary factor in the "$\mathbf{QR}$" decomposition of $\mathbf{A}^n$ then

$$\mathbf{A}_{n+1}=\mathbf{P}_n^*\mathbf{A}\,\mathbf{P}_n, \tag{7}$$

which means that convergence of $\{\mathbf{A}_n\}$ follows from convergence of $\{\mathbf{P}_n\}$. (All this is true in the case of $m\times m$ matrices $\mathbf{A}$.)

It is easy to write down explicitly the "$\mathbf{QR}$" decomposition of a 2×2 matrix. If

$$\mathbf{A}=\mathbf{A}_1=\begin{bmatrix}\alpha & \beta\\ \gamma & \delta\end{bmatrix}=\begin{bmatrix}c & s\\ s & -c\end{bmatrix}\begin{bmatrix}a & b\\ 0 & c\end{bmatrix}$$

then

$$c=\alpha/\sqrt{\alpha^2+\gamma^2}, \quad s=\gamma/\sqrt{\alpha^2+\gamma^2}. \tag{8}$$

From (5) we have, assuming $\mathbf{A}$ unimodular, i.e. $\alpha\delta-\beta\gamma=1$,

$$\mathbf{A}_2=\begin{bmatrix}c & s\\ s & -c\end{bmatrix}\begin{bmatrix}\alpha & \beta\\ \gamma & \delta\end{bmatrix}\begin{bmatrix}c & s\\ s & -c\end{bmatrix}=\begin{bmatrix}\alpha+\delta-(\alpha/(\alpha^2+\gamma^2)) & \gamma-\beta-(\gamma/(\alpha^2+\gamma^2))\\ -\gamma/(\alpha^2+\gamma^2) & \alpha/(\alpha^2+\gamma^2)\end{bmatrix}.$$

If we now write, for $n=1,2,\ldots$, $\mathbf{A}_n=\begin{bmatrix}\alpha_n & \beta_n\\ \gamma_n & \delta_n\end{bmatrix}$ the recurrence relations for the $\alpha_n, \beta_n, \ldots$ can be written down. In particular,

$$\begin{aligned}\alpha_{n+1}&=(\alpha_1+\delta_1)-(\alpha_n/(\alpha_n^2+\gamma_n^2))\\ \gamma_{n+1}&=-\gamma_n/(\alpha_n^2+\gamma_n^2).\end{aligned} \tag{9}$$

What we want to prove is that, in certain circumstances, if λ is an appropriate characteristic value of $\mathbf{A}_1$, then

$$\gamma_n\to 0, \quad \alpha_n\to\lambda, \quad \delta_n\to\lambda^{-1}.$$

The general solution of the non-linear recurrence relations, (9), is not simple — we discuss a special case first.

We take the matrix $\begin{bmatrix} 2 & -1 \\ -1 & 2 \end{bmatrix}$ which we must normalize to $\mathbf{A}_1 = \begin{bmatrix} 2/\sqrt{3} & -1/\sqrt{3} \\ -1/\sqrt{3} & 2/\sqrt{3} \end{bmatrix}$. This has characteristic roots $\sqrt{3}$, $1/\sqrt{3}$. Application of the recurrence relations (9) gives

	$n=1$	$n=2$	$n=3$	$n=4$
$\alpha_n/\sqrt{3}=$	$2/3$	$14/15$	$122/123$	$1094/1095$
$\gamma_n/\sqrt{3}=$	$-1/3$	$1/5$	$-3/41$	$9/365$

which indicates that $\alpha_n \to \sqrt{3}$, $\gamma_n \to 0$. The general form of α_n, γ_n can be conjectured from the above table and established by induction. We find

$$(10) \qquad (\alpha_n/\sqrt{3}) = \frac{1+3\times 9^{n-1}}{3(1+9^{n-1})}, \quad (-1)^n (\gamma_n/\sqrt{3}) = \frac{2\times 3^{n-1}}{3(1+9^{n-1})}.$$

These results indicate that the convergence of γ_n is ultimately geometric with common ratio $\frac{1}{3}=\lambda^{-2}$ while that of α_n is ultimately geometric with common ratio $\frac{1}{9}=\lambda^{-4}$.

The general real 2×2 case can be handled in the following way. (Cf. E. Kreyszig and J. Todd, Elem. Math., to appear.) In (9) we write $x_n=\alpha_n$, $y_n=(-1)^n\gamma_n \alpha_1+\delta_1=k$ and get

$$(11) \qquad \begin{aligned} x_{n+1} &= k - x_n/(x_n^2+y_n^2) \\ y_{n+1} &= y_n/(x_n^2+y_n^2), \end{aligned}$$

which we combine as

$$z_{n+1} = k - \bar{z}_n/(z_n \bar{z}_n)$$

i.e.,

$$z_{n+1} = k - z_n^{-1}, \quad n=1, 2, \dots,$$

where $z_r = x_r + iy_r$.

We have therefore reduced our problem to that of the iteration of the bilinear transformation

$$w = \frac{kz-1}{z}.$$

It is easy to check that if

$$W = \frac{\alpha w+\beta}{\gamma w+\delta}, \quad w = \frac{az+b}{cz+d}$$

then

$$W=\frac{Az+B}{Cz+D} \quad \text{where} \quad \begin{bmatrix} A & B \\ C & D \end{bmatrix} = \begin{bmatrix} \alpha & \beta \\ \gamma & \delta \end{bmatrix} \begin{bmatrix} a & b \\ c & d \end{bmatrix}.$$

Thus the iteration of a bilinear transformation is equivalent to the powering of a matrix.

We use the following fact (Problem 7.5).

If $\mathbf{A}$ is a 2×2 matrix with distinct characteristic values λ, μ then

$$\mathbf{A}^n=\begin{bmatrix} ad\lambda^n-bc\mu^n & -ab(\lambda^n-\mu^n) \\ cd(\lambda^n-\mu^n) & -bc\gamma^n+ad\mu^n \end{bmatrix} \tag{12}$$

where

$$\mathbf{T}=\begin{bmatrix} a & b \\ c & d \end{bmatrix}$$

is an unimodular matrix which diagonalizes $\mathbf{A}$, i.e.,

$$\mathbf{T}^{-1}\mathbf{AT}=\operatorname{diag}[\lambda, \mu].$$

The matrix $\mathbf{A}=\begin{bmatrix} k & -1 \\ 1 & 0 \end{bmatrix}$, where we now assume $k>2$, is diagonalized by

$$\mathbf{T}=\frac{1}{\sqrt{\lambda-\lambda^{-1}}}\begin{bmatrix} \lambda & \lambda^{-1} \\ 1 & 1 \end{bmatrix},$$

where $\lambda(>1)$ and λ^{-1} are the characteristic values of $\mathbf{A}$ and also of $\mathbf{A}_1= \begin{bmatrix} \alpha_1 & \beta_1 \\ \gamma_1 & \delta_1 \end{bmatrix}$.

The rest of the discussion is a matter of elementary algebra.

Writing $v_n=\lambda^n-\lambda^{-n}$ we find from (12)

$$z_{n+1} \equiv x_{n+1}+iy_{n+1} = \frac{v_{n+1}(x_1+iy_1)-v_n}{v_n(x_1+iy_1)-v_{n-1}}.$$

Multiplying above and below on the right by $v_n(x_1-iy_1)-v_{n-1}$ and equating real and imaginary parts shows that

$$x_{n+1}=[v_n v_{n+1}(x_1^2+y_1^2)+v_n v_{n-1}-x_1(v_n^2+v_{n+1}v_{n-1})]/D,$$

$$y_{n+1}=(v_n^2-v_{n+1}v_{n-1})y_1/D,$$

where

$$D=v_n^2(x_1^2+y_1^2)-2v_{n-1}v_n x_1+v_{n-1}^2.$$

We find

$$x_{n+1}-\lambda=[2(\lambda-\lambda^{-1})\{x_1^2+y_1^2+1-x_1(\lambda+\lambda^{-1})\}+\mathcal{O}(\lambda^{-2n})]/D,$$

$$y_{n+1}=(\lambda-\lambda^{-1})^2 y_1/D,$$

where

$$D=[(x_1-\lambda^{-1})^2+y_1^2]\lambda^{2n}+\mathcal{O}(1).$$

These results check with those given in (10) for the special case. Note that when the matrix $\mathbf{A}_1$ is symmetric, as well as unimodular, we have $\alpha_1\delta_1-\gamma_1^2=1$, i.e., $\alpha_1(k-\alpha_1)-\gamma_1^2=1$, i.e., $\alpha_1^2+\gamma_1^2+1=k\alpha_1$ so that we have $x_n-\lambda=\mathcal{O}(\lambda^{-4n})$, $y_n=\mathcal{O}(\lambda^{-2n})$.

For further information about this method see the original papers of Rutishauser, Francis and Kublanowskaja, the book of Wilkinson and various expository articles, e.g., those of Parlett.

Chapter 8, Problems

8.1. Show that trigonometrical tables are not essential for the carrying out of Jacobi rotations, square root tables sufficing.

8.2. Write a program for the Jacobi method and apply it to some of the symmetric matrices which we have been studying.

8.3.

		0	
0			

		0	0
0			
0			

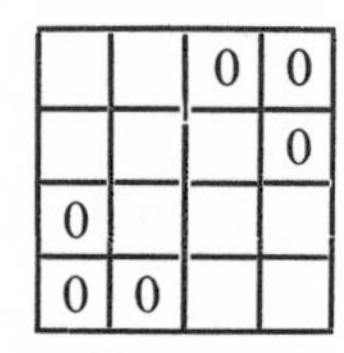

Indicate by an * those elements which can change at each stage in the Givens reduction of a 4×4 symmetric matrix.

8.4. Write a program for carrying out the reduction to triple diagonal form for a small matrix and carry it out in some special cases.

8.5. Write a program to evaluate a polynomial $p(x)$ at the points $a+\frac{p}{2^q}(b-a)$ where a, b are given and $p=0, 1, \ldots, 2^q$. Develop this program to give as an output a "graph" of $p(x)$.

8.6. A theorem of Sturm (1803—1855) is as follows: if $p_0(x)$ is a polynomial of degree n with real coefficients and no multiple roots, if $p_1=p_0'$ and if we define $p_2, p_3, \ldots, p_n$ by the relations

$$p_{r-1}(x)=q_r(x)p_r(x)-p_{r+1}(x) \quad r=1, 2, \ldots, n-1$$

obtained by "long division" then the number of real zeros of $p_0(x)$ between a, b is exactly $\Delta(a)-\Delta(b)$ where

$$\Delta(x)=\delta(p_0(x), p_1(x), \ldots, p_n(x))$$

where δ is defined in Problem 1.11.

Check this theorem in some simple cases. Then write a program to use it for a polynomial of degree 6, say. (Note that in "practical computation" you will have to check "$|a| \gtrless t$", where t is some tolerance.)

Check your program by locating approximately the real zeros of

$$2x^6+15x^5-98x^4-281x^3+693x^2+742x-1330$$

and of

$$x^6-6x^5-30x^2+12x-9.$$

8.7. Show how to evaluate $f_n(\lambda)$ and $f_n'(\lambda)$ for specific λ, where $f_n(\lambda)$ is the characteristic polynomial of the triple diagonal $n \times n$ matrix

$$\mathbf{A}=[\ldots, \overset{\cdots}{\underset{\cdots}{c_r, a_r, b_r}}, \ldots].$$

Do the same when $f_n(\lambda)$ is the characteristic polynomial of a Hessenberg matrix.

Use your program to evaluate the characteristic polynomials of

$$\mathbf{A}=\begin{bmatrix} 0 & 10 & 0 & 0 & 0 \\ 10 & 0 & 10 & 0 & 0 \\ 0 & 10 & 0 & 10 & 0 \\ 0 & 0 & 10 & 0 & 10 \\ 0 & 0 & 0 & 10 & 0 \end{bmatrix}, \quad \mathbf{B}=\begin{bmatrix} 0 & 1 & 0 & 0 & 0 \\ 3 & 0 & 1 & 0 & 0 \\ 4 & 3 & 0 & 1 & 0 \\ 5 & 4 & 3 & 0 & 1 \\ 6 & 5 & 4 & 3 & 0 \end{bmatrix}$$

in the interesting ranges.

8.8. Using the results of Problems 8.6, 8.7, draw a flow-diagram for a program to compute the characteristic values of a symmetric triple diagonal matrix.

Indicate any modifications required when the matrix is in Hessenberg form. (See Problem 4.12.)

8.9. Apply the Givens and the Householder method to reduce the matrix

$$\begin{bmatrix} 5 & 7 & 6 & 5 \\ 7 & 10 & 8 & 7 \\ 6 & 8 & 10 & 9 \\ 5 & 7 & 9 & 10 \end{bmatrix}$$

to triple diagonal form. The result obtained by one program is

$$\mathbf{W}_1=\begin{bmatrix} 5 & 10.488089 & 0 & 0 \\ 10.488089 & 25.472729 & 3.521903 & 0 \\ 0 & 3.521898 & 3.680571 & -.185813 \\ 0 & 0 & -.185813 & .846701 \end{bmatrix}.$$

Theoretically this should be symmetric, and it is unnecessary to calculate the elements below the diagonal. However, if these are calculated, the differences in symmetric elements gives some idea of the errors occurring.

Find the characteristic values of $\mathbf{W}_1$, e.g., by drawing a rough graph of the characteristic polynomial of $\mathbf{W}_1$, and then using Newton's method to estimate the characteristic values more accurately.

8.10. In the so-called "cyclic-Jacobi" method we kill the elements off in order

$$a_{12}, a_{13}, \ldots, a_{1n};\ a_{23}, a_{24}, \ldots, a_{2n};\ \ldots;\ a_{n-1,n}$$

and then go back to the start of the cycle and repeat.

Carry out two full cycles (i.e., 6 rotations) in the case of the matrix

$$\mathbf{A}=\begin{bmatrix} 0 & 1 & 1 \\ 1 & 4 & 0 \\ 1 & 0 & 8 \end{bmatrix}.$$

8.11. Show that the tridiagonalization of a symmetric matrix can also be accomplished by the Rutishauser process:
annihilate the elements in $(1, n)$, $(1, n-1)$, ..., $(1, 3)$; $(2, n)$, $(2, n-1)$, ..., $(2, 4)$; ... positions by rotations in the $(n-1, n)$, $(n-2, n-1)$, ..., $(2, 3)$; $(n-1, n)$...$(3, 4)$; ... planes.

8.12. Find the characteristic values of the triple diagonal matrix

$$\begin{bmatrix} 0 & 5 & 0 & 0 & 0 & 0 \\ 1 & 0 & 4 & 0 & 0 & 0 \\ 0 & 1 & 0 & 3 & 0 & 0 \\ 0 & 0 & 1 & 0 & 2 & 0 \\ 0 & 0 & 0 & 1 & 0 & 1 \\ 0 & 0 & 0 & 0 & 1 & 0 \end{bmatrix}.$$

8.13. (a) Find a rough approximation $\tilde{\alpha}$ to the dominant characteristic value α of

$$\mathbf{A}=\begin{bmatrix} 5 & 4 & 3 & 2 & 1 \\ 4 & 6 & 0 & 4 & 3 \\ 3 & 0 & 7 & 6 & 5 \\ 2 & 4 & 6 & 8 & 7 \\ 1 & 3 & 5 & 7 & 9 \end{bmatrix}.$$

(b) Use Wielandt inverse iteration to find the dominant characteristic vector of $\mathbf{A}$. Use the Rayleigh quotient to get an improved estimate for α.

(c) Find the characteristic vector of $\mathbf{A}$ corresponding to its characteristic root near 5.

8.14. Show how to represent a matrix $\mathbf{A}$ in the form $\mathbf{A}=\mathbf{Q}\mathbf{R}$, where $\mathbf{Q}$ is unitary and $\mathbf{R}$ upper triangular by the use of Householder transformations.

8.15. Find all the characteristic values of $\mathbf{A}_{11}$ of Problem 1.2 in the case $n=6$.

CHAPTER 9

Iterative Methods for the Solution of Systems $\boldsymbol{A}\boldsymbol{x}=\boldsymbol{b}$

The general idea here is to guess an answer $\boldsymbol{x}^{(0)}$, the nearer to $\boldsymbol{x}$ the better, and to use some iterative process which gets cheaply from any $\boldsymbol{x}^{(r)}$ a presumably better $\boldsymbol{x}^{(r+1)}$ and, at any rate, we are to have

$$\boldsymbol{x}^{(r)} \to \boldsymbol{x}.$$

1. The Jacobi and Gauss—Seidel Methods

We illustrate this in two classical cases. The first is the naive Jacobi process. Given $\boldsymbol{x}^{(r)}=[x_1^{(r)}, \ldots, x_n^{(r)}]'$ we get the next approximation as follows:

(1) Substitute $x_2^{(r)}, \ldots, x_n^{(r)}$ for $x_2, \ldots, x_n$ in the first equation and solve for x_1 which we define as $x_1^{(r+1)}$.

(2) Substitute $x_1^{(r)}, x_3^{(r)}, \ldots, x_n^{(r)}$ for $x_1, x_3, \ldots, x_n$ in the second equation and solve for x_2 which we define as $x_2^{(r+1)}$.

...

(n) Substitute $x_1^{(r)}, x_2^{(r)}, \ldots, x_{n-1}^{(r)}$ for $x_1, x_2, \ldots, x_{n-1}$ in the n-th equation and solve for x_n which we define as $x_n^{(r+1)}$.

This process can only be carried out if all the diagonal elements of $\boldsymbol{A}$ are non-zero.

Consider the following system

$$\begin{aligned} 12x-3y+2z&=96 \\ -3x-8y+\ z&=68 \\ x+2y+6z&=3 \end{aligned} \qquad (1)$$

for which the exact solution is

$$x=5, \quad y=-10, \quad z=3.$$

If we choose $\boldsymbol{x}^{(0)}=\boldsymbol{e}$ the successive approximations to the solution are, working to $2D$:

$$\begin{bmatrix}1\\1\\1\end{bmatrix}, \begin{bmatrix}8.08\\-8.75\\0\end{bmatrix}, \begin{bmatrix}5.81\\-11.53\\1.74\end{bmatrix}, \ldots, \boldsymbol{x}^{(10)}=\begin{bmatrix}\ \\ \ \\ \ \end{bmatrix}, \ldots, \boldsymbol{x}^{(20)}=\begin{bmatrix}\ \\ \ \\ \ \end{bmatrix}.$$

Convergence seems to be in progress, but it is not very rapid.

It is not difficult to verify that if we assume diag $\boldsymbol{A}=\boldsymbol{e}$ then this algorithm can be described as

$$\boldsymbol{x}^{(r+1)}=(\boldsymbol{A}-\boldsymbol{I})\boldsymbol{x}^{(r)}+\boldsymbol{b}.$$

Assuming that $\boldsymbol{A}$ is non-singular, so that there is a solution $\boldsymbol{x}$, which satisfies $\boldsymbol{A}\boldsymbol{x}=\boldsymbol{b}$ which can be written as

$$\boldsymbol{x}=(\boldsymbol{I}-\boldsymbol{A})\boldsymbol{x}+\boldsymbol{b},$$

we can find a recurrence relation for errors $\boldsymbol{\varepsilon}^{(r)}=\boldsymbol{x}^{(r)}-\boldsymbol{x}$. It is got by subtracting the last two equations

$$\boldsymbol{\varepsilon}^{(r+1)}=(\boldsymbol{I}-\boldsymbol{A})\boldsymbol{\varepsilon}^{(r)}.$$

In this form we find it easy to analyse the Jacobi process. We would like to have $\boldsymbol{x}^{(r)}\to\boldsymbol{x}$ no matter what $\boldsymbol{x}^{(0)}$ is, i.e. $\boldsymbol{\varepsilon}^{(r)}\to 0$ no matter what $\boldsymbol{\varepsilon}^{(0)}$ may be.

Suppose that the matrix $\boldsymbol{A}$ has a full set of characteristic vectors $\boldsymbol{c}_i$ corresponding to characteristic values α_i, $i=1, 2, \ldots, n$. Let us expand $\boldsymbol{\varepsilon}^{(0)}$ in terms of the $\boldsymbol{c}_i$'s as

$$\boldsymbol{\varepsilon}^{(0)}=\sum a_i \boldsymbol{c}_i.$$

Then

$$\boldsymbol{\varepsilon}^{(r)}=(\boldsymbol{I}-\boldsymbol{A})^r\boldsymbol{\varepsilon}^{(0)}$$

$$=\sum(1-\alpha_i)^r a_i \boldsymbol{c}_i$$

and so

$$\boldsymbol{\varepsilon}^{(r)}\to 0$$

or all $\boldsymbol{\varepsilon}^{(0)}$, if $|1-\alpha_i|<1$ for all i, i.e., if the spectral radius of $\boldsymbol{I}-\boldsymbol{A}$ is less than 1.

An alternative analysis of this process is the following. We assume a vector norm given and a matrix norm which is the induced norm, or is compatible with the vector norm. Then, by submultiplicativity

$$\|\boldsymbol{\varepsilon}^{(r)}\|\leqq(\|\boldsymbol{I}-\boldsymbol{A}\|)^r\|\boldsymbol{\varepsilon}^{(0)}\|$$

and if $\|\boldsymbol{I}-\boldsymbol{A}\|=\theta<1$ we certainly have convergence, with no assumption on the characteristic vectors of $\boldsymbol{A}$. This result includes the previous one for it can be shown that for any matrix $\boldsymbol{A}$, there is a matrix norm for which $\|\boldsymbol{A}\|<\varrho(\boldsymbol{A})+\varepsilon$, for any $\varepsilon>0$. We have discussed this statement in Chapter 3. Normalising the system just discussed we get

$$\boldsymbol{A}=\begin{bmatrix}1, & -1/4, & 1/6\\ 3/8, & 1, & -1/8\\ 1/6, & 1/3, & 1\end{bmatrix}$$

so that the Gerschgorin configuration for $\boldsymbol{I}-\boldsymbol{A}$ consists of two circles (center the origin) with radius 1/2 and one circle (center the origin) with 5/12. Hence the spectral radius of $\boldsymbol{I}-\boldsymbol{A}$ is certainly $\leqq\frac{1}{2}<1$ and convergence is assured.

Observe that if we do not insist on the normalization of the system our condition will be

$$\varrho(\mathbf{I}-\mathbf{D}^{-1}\mathbf{A})<1$$

where $\mathbf{D}=\operatorname{diag}\mathbf{A}$. We have already noted that the process is applicable only when $\mathbf{D}$ is non-singular.

We next discuss the Gauss—Seidel process. It seems wasteful in the Jacobi process that we do not make use of the best estimates on hand for the x_i's. This means that we proceed as follows:

(1) as in the Jacobi case;

(2) substitute the new estimate $x_1^{(r+1)}$ for x_1 and the old estimates $x_3^{(r)}, \dots, x_n^{(r)}$ for $x_3, \dots, x_n$ in the second equation and solve for x_2 which we define as $x_2^{(r+1)}$.

...

(n) substitute $x_1^{(r+1)}, \dots, x_{n-1}^{(r+1)}$ for $x_1, \dots, x_{n-1}$ in the n-th equation and solve for x_n which we define as $x_n^{(r+1)}$.

If we apply this to the system (1), beginning with the same $\mathbf{x}^{(0)}=\mathbf{e}$ the successive approximations to the solution, are, working to $2D$:

$$\begin{bmatrix}1\\1\\1\end{bmatrix},\ \begin{bmatrix}8.08\\-11.40\\2.95\end{bmatrix},\ \begin{bmatrix}4.66\\-9.88\\3.02\end{bmatrix},\ \dots,\ \begin{bmatrix}5.03\\10.01\\3.00\end{bmatrix},\ \dots$$

so that more rapid convergence is apparent.

With the normalizing assumption $\operatorname{diag}\mathbf{A}=\mathbf{e}$ and writing

$$\mathbf{A}=-\mathbf{L}+\mathbf{I}-\mathbf{U}$$

where $\mathbf{L}(\mathbf{U})$ is strictly lower (upper) triangular we can describe the algorithm by the equation

$$(\mathbf{I}-\mathbf{L})\,\mathbf{x}^{(r+1)}=\mathbf{b}+\mathbf{U}\,\mathbf{x}^{(r)}.$$

To clarify the derivation of this basic relation we give the full details in the case $n=3$, with a different notation:

$$\left.\begin{aligned} x+b_1y+c_1z&=d_1\\ a_2x+\quad y+c_2z&=d_2\\ a_3x+b_3y+\quad z&=d_3\end{aligned}\right\}$$

$$\left.\begin{aligned} x^{(r+1)}&=d_1-b_1y^{(r)}-c_1z^{(r)}\\ y^{(r+1)}&=d_2\qquad\qquad -c_2z^{(r)}-a_2x^{(r+1)}\\ z^{(r+1)}&=d_3\qquad\qquad\qquad\qquad -a_3x^{(r+1)}-b_3y^{(r+1)}\end{aligned}\right\}$$

$$\begin{bmatrix}x^{(r+1)}\\y^{(r+1)}\\z^{(r+1)}\end{bmatrix}=\begin{bmatrix}d_1\\d_2\\d_3\end{bmatrix}-\begin{bmatrix}0&b_1&c_1\\0&0&c_2\\0&0&0\end{bmatrix}\begin{bmatrix}x^{(r)}\\y^{(r)}\\z^{(r)}\end{bmatrix}-\begin{bmatrix}0&0&0\\a_2&0&0\\a_3&b_3&0\end{bmatrix}\begin{bmatrix}x^{(r+1)}\\y^{(r+1)}\\z^{(r+1)}\end{bmatrix}.$$

Then subtracting from this equality the original equation which can be written in the form

$$(\boldsymbol{I}-\boldsymbol{L})\boldsymbol{x}=\boldsymbol{b}+\boldsymbol{U}\boldsymbol{x}$$

we get

$$(\boldsymbol{I}-\boldsymbol{L})\,\varepsilon^{(r+1)}=\boldsymbol{U}\,\varepsilon^{(r)}$$

which gives

$$\varepsilon^{(r+1)}=(\boldsymbol{I}-\boldsymbol{L})^{-1}\boldsymbol{U}\,\varepsilon^{(r)}$$

so that

$$\varepsilon^{(r)}=\{(\boldsymbol{I}-\boldsymbol{L})^{-1}\boldsymbol{U}\}^{r}\,\varepsilon^{(0)}.$$

We see, as before, that a necessary and sufficient condition for convergence is that the spectral radius of $(\boldsymbol{I}-\boldsymbol{L})^{-1}\boldsymbol{U}$ should be less than unity. In the special case we have

$$(\boldsymbol{I}-\boldsymbol{L})=\begin{bmatrix}1 & 0 & 0\\ 3/8 & 1 & 0\\ 1/6 & 1/3 & 1\end{bmatrix},\quad (\boldsymbol{I}-\boldsymbol{L})^{-1}=\begin{bmatrix}1 & 0 & 0\\ -3/8 & 1 & 0\\ -1/24 & -1/3 & 1\end{bmatrix}$$

$$\boldsymbol{U}=\begin{bmatrix}0 & 1/4 & -1/6\\ 0 & 0 & -1/8\\ 0 & 0 & 0\end{bmatrix}$$

so that

$$(\boldsymbol{I}-\boldsymbol{L})^{-1}\boldsymbol{U}=\begin{bmatrix}0 & 1/4 & -1/6\\ 0 & -3/32 & 3/16\\ 0 & -1/36 & -5/144\end{bmatrix}.$$

This has spectral radius at most 5/12, by Gerschgorin.

This criterion is not a very practical one: it is awkward to determine in general the characteristic values of the product of two matrices in terms of those of the factors. (This is easy, of course, if they commute.)

A different discussion shows that convergence takes place if $\boldsymbol{A}$ is a positive definite symmetric matrix. (See Problem 9.8.)

It is plausible to think that the Gauss—Seidel method would be better than the Jacobi one. Actually this is not so, and examples can be constructed to show that the methods are not comparable. For some matrices $\boldsymbol{A}$ the spectral radius of $(\boldsymbol{I}-\boldsymbol{A})$ is less than that of $(\boldsymbol{I}-\boldsymbol{L})^{-1}\boldsymbol{U}$, and for others the reverse is the case. (See Problem 9.2.)

When are these iterative methods likely to be superior to the direct Gaussian method? One promising case is when $\boldsymbol{A}$ is "sparse", i.e. most of the a_{ij} are zero, e.g. when $\boldsymbol{A}$ is triple diagonal. Let us investigate the Jacobi case when $\boldsymbol{A}=\boldsymbol{A}_8$.

We have to normalize $\boldsymbol{A}_8$ to get a matrix $\tilde{\boldsymbol{A}}$ with a unit diagonal. The characteristic values of $\tilde{\boldsymbol{A}}$ are (see Problem 5.14 (iv)):

$$\tilde{\alpha}_k = 2\sin^2 k\pi/2(n+1) \quad k=1,2,\ldots,n.$$

The characteristic values of $\boldsymbol{I}-\tilde{\boldsymbol{A}}$ are

$$1-2\sin^2 k\pi/2(n+1)$$

and these lie between

$$\pm(1-2\sin^2 \pi/2(n+1)) \sim \pm(1-\pi^2/2n^2)$$

so that $\varrho(\boldsymbol{I}-\tilde{\boldsymbol{A}})<1$ and the Jacobi process is convergent.

The spectral radius gives an upper bound for the decrease in $\|\varepsilon^{(r)}\|$ per cycle. Suppose we carry out n^2 Jacobi cycles. Then an estimate for the total attenuation is

$$(1-\pi^2/2n^2)^{n^2} \doteqdot e^{-\pi^2/2} \doteqdot 0.007.$$

Hence after $4n^2$ cycles we would have an attenuation of more than 10^{-8}. Thus we might expect to obtain useful results after $4n^2$ cycles. Now if the matrix were full, each cycle would involve about n^2 multiplications and in all $\mathcal{O}(n^4)$ which is not competitive with the direct method. If the matrix were "sparse", e.g. triple diagonal, each cycle would involve $\mathcal{O}(n)$ multiplications and in all $\mathcal{O}(n^3)$ which is to be compared with $\mathcal{O}(n)$ multiplications needed in an efficient direct handling of this case.

2. Young overrelaxation method

It therefore appears that more powerful methods than the Jacobi method must be sought. The Gauss—Seidel is not much improvement, the spectral radius of the iteration matrix is still $1-\mathcal{O}(n^{-2})$ in practical cases. However in an important practical case the Young Successive Overrelaxation Method has a spectral radius of $1-\mathcal{O}(1/n)$ and the Alternating Direction Implicit Method has a spectral radius $1-\mathcal{O}(1/\log n)$ and these are quite feasible.

We cannot discuss these methods fully here and restrict our considerations to the use of the Young scheme on a special 4×4 system: $\boldsymbol{A}\boldsymbol{x}=\boldsymbol{b}$ where

$$\boldsymbol{A}=\begin{bmatrix} 1 & 0 & -\frac{1}{4} & -\frac{1}{4} \\ 0 & 1 & -\frac{1}{4} & -\frac{1}{4} \\ -\frac{1}{4} & -\frac{1}{4} & 1 & 0 \\ -\frac{1}{4} & -\frac{1}{4} & 0 & 1 \end{bmatrix} = \boldsymbol{I}-\boldsymbol{L}-\boldsymbol{U}$$

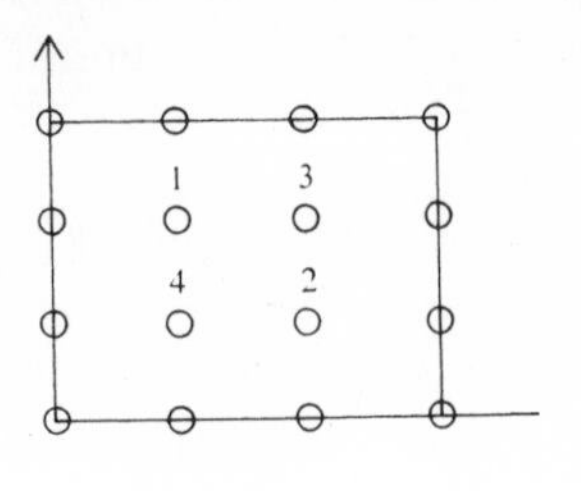

which arises from the approximate solution to the Laplace equation

$$V_{xx}+V_{yy}=0$$

in a unit square where V is given on the boundary. In fact, x_1, x_2, x_3, x_4 are approximations to $V(\frac{1}{3}, \frac{2}{3})$, $V(\frac{2}{3}, \frac{1}{3})$, $V(\frac{2}{3}, \frac{2}{3})$, $V(\frac{1}{3}, \frac{1}{3})$ where b_1, b_2, b_3, b_4 are appropriate linear combinations of the boundary values.

The Gerschgorin circles for $\boldsymbol{A}$ all have center 1 and radius $\frac{1}{2}$. Hence $\boldsymbol{A}$ is non-singular and indeed positive definite. It can be verified that

$$6\boldsymbol{A}^{-1}=\begin{bmatrix}7 & 1 & 2 & 2\\ 1 & 7 & 2 & 2\\ 2 & 2 & 7 & 1\\ 2 & 2 & 1 & 7\end{bmatrix}.$$

Since $\boldsymbol{A}$ is symmetric, the α's are real and, by Gerschgorin,

$$\tfrac{1}{2}\leqq\alpha_i\leqq\tfrac{3}{2}$$

which gives

$$-\tfrac{1}{2}\leqq 1-\alpha_i\leqq\tfrac{1}{2}$$

so that $\varrho(\boldsymbol{I}-\boldsymbol{A})\leqq\frac{1}{2}$ and the Jacobi process converges. It can be shown (using Problem 4.10, e.g.,) that the actual characteristic values of $\boldsymbol{I}-\boldsymbol{A}$ are

$$0,\ 0,\ 1/2,\ -1/2.$$

To discuss the Gauss—Seidel process in this case we denote by $\boldsymbol{J}$ the matrix $\begin{bmatrix}1 & 1\\ 1 & 1\end{bmatrix}$ and compute

$$[\boldsymbol{I}-\boldsymbol{L}]^{-1}=\begin{bmatrix}\boldsymbol{I} & 0\\ -\frac{1}{4}\boldsymbol{J} & \boldsymbol{I}\end{bmatrix}^{-1}=\begin{bmatrix}\boldsymbol{I} & 0\\ \frac{1}{4}\boldsymbol{J} & \boldsymbol{I}\end{bmatrix},$$

and

$$[\boldsymbol{I}-\boldsymbol{L}]^{-1}\boldsymbol{U}=\begin{bmatrix}\boldsymbol{I} & 0\\ \frac{1}{4}\boldsymbol{J} & \boldsymbol{I}\end{bmatrix}\begin{bmatrix}0 & \frac{1}{4}\boldsymbol{J}\\ 0 & 0\end{bmatrix}=\begin{bmatrix}0 & \frac{1}{4}\boldsymbol{J}\\ 0 & \frac{1}{16}\boldsymbol{J}^2\end{bmatrix}=\begin{bmatrix}0 & \frac{1}{4}\boldsymbol{J}\\ 0 & \frac{1}{8}\boldsymbol{J}\end{bmatrix},$$

since $\boldsymbol{J}^2=2\boldsymbol{J}$. The characteristic values of $(\boldsymbol{I}-\boldsymbol{L})^{-1}\boldsymbol{U}$ are therefore

$$0,\ 0,\ 0,\ \tfrac{1}{4}.$$

The spectral radius of the error matrix is $\frac{1}{2}$ in the Jacobi case and $\frac{1}{4}$ in the Gauss—Seidel case.

Referring back to our discussion of the Gauss—Seidel process we see that in it we replace $x_s^{(r)}$ by

$$x_s^{(r+1)}=b_s-[a_{s,1}x_1^{(r+1)}+\ldots+a_{s,s-1}x_{s-1}^{(r+1)}]-[a_{s,s+1}x_{s+1}^{(r)}+\ldots+a_{s,n}x_n^{(r)}],$$

i.e., we add to x_s the quantity

$$b_s-[\ldots]-[\ldots]-x_s.$$

It is plausible that "over-correcting" might improve convergence, and so we consider replacing $x_s^{(r)}$ by $x_s^{(r+1)}=x_s+\omega\{b_s-[\ldots]-[\ldots]-x_s\}$ where ω is a parameter which we shall optimize. We can express this change in matrix language as:

$$(\boldsymbol{I}-\omega\boldsymbol{L})\boldsymbol{x}^{(r+1)}=\{(1-\omega)\boldsymbol{I}+\omega\boldsymbol{U}\}\boldsymbol{x}^{(r)}+\omega\boldsymbol{b}$$

and subtracting from this the relation $\omega\boldsymbol{A}\boldsymbol{x}=\omega\boldsymbol{b}$ written in the form

$$(\boldsymbol{I}-\omega\boldsymbol{L})\boldsymbol{x}=\{(1-\omega)\boldsymbol{I}+\omega\boldsymbol{U}\}\boldsymbol{x}+\omega\boldsymbol{b}$$

we get

$$\varepsilon^{(r+1)}=(\boldsymbol{I}-\omega\boldsymbol{L})^{-1}\{(1-\omega)\boldsymbol{I}+\omega\boldsymbol{U}\}\varepsilon^{(r)},\quad \text{say}\quad \varepsilon^{(r+1)}=\mathcal{L}_\omega\varepsilon^{(r)}.$$

For $\omega=1$ this checks with the Gauss—Seidel relation.

Our object now is to determine the spectral radius of $\mathcal{L}_\omega$ and then choose ω to make it least.

We shall only discuss this in the special case. It is easy to find

$$(\boldsymbol{I}-\omega\boldsymbol{L})^{-1}=\begin{bmatrix}\boldsymbol{I} & 0\\ \tau\boldsymbol{J} & \boldsymbol{I}\end{bmatrix},\quad \text{where}\quad \tau=\omega/4,$$

and that

$$\mathcal{L}_\omega=\begin{bmatrix}\sigma\boldsymbol{I} & \tau\boldsymbol{J}\\ \sigma\tau\boldsymbol{J} & 2\tau^2\boldsymbol{J}+\alpha\boldsymbol{I}\end{bmatrix},\quad \text{where}\quad \sigma=1-\omega.$$

We can write

$$\mathcal{L}_\omega=\sigma\boldsymbol{I}=\begin{bmatrix}0 & \tau\boldsymbol{J}\\ \sigma\tau\boldsymbol{J} & 2\tau^2\boldsymbol{J}\end{bmatrix}$$

and we *find* (cf. Problem 9.6) that the characteristic values of $\mathcal{L}_\omega$ are

$$\sigma,\ \sigma,\ \sigma+\left\{2\tau^2\pm2\sqrt{\tau^4+\sigma\tau^2}\right\},$$

i.e.,

$$1-\omega,\ 1-\omega\quad\text{and}\quad 1-\omega+\left\{\omega^2\pm\sqrt{\omega^4+16(1-\omega)\omega^2}\right\}/8.$$

It is easy to compute the spectral radius ϱ_ω for different values of ω. We find

$$\varrho_0=1,\quad \varrho_{1/2}=\frac{17+\sqrt{33}}{32}\sim0.71,\quad \varrho_1=\frac{1}{4},\quad \varrho_2=1.$$

The least value of ϱ_ω occurs when

$$\omega = 8 - \sqrt{48} \doteqdot 1.07$$

and is

$$\varrho_b = 7 - \sqrt{48} \doteqdot 0.07.$$

Thus the spectral radius of the error matrix in the Young case, with optimal parameter, is $7-\sqrt{48}$ which we compare with $\frac{1}{2}$ in the Jacobi case and $\frac{1}{4}$ in the Gauss—Seidel case. Thus the Young method is about four times as fast as the Gauss—Seidel method. (Cf. Problem 9.6.)

3. Gradient or steepest descent methods

These form another class of iterative methods of considerable interest. We begin with a simple example:

$$\left.\begin{aligned} x - y &= 0 \\ -x + 2y &= 1 \end{aligned}\right\}.$$

The solution to this system is the point which minimizes the quadratic form

$$\varepsilon(x, y) = (x-y)^2 + (-x+2y-1)^2.$$

We want to show how to construct a sequence of points (x_n, y_n), the first of which we take as $(x_0, y_0) = (0, 0)$, which converge to the solution (1, 1).

Consider the curves

$$\varepsilon(x, y) = \text{constant}. \tag{2}$$

These form a family of concentric ellipses centered at (1, 1) and whose axes are the lines

$$(y-1) = \left(\frac{-1 \pm \sqrt{5}}{2}\right)(x-1).$$

We want to proceed from (x_0, y_0) in the direction in which $\varepsilon(x, y)$ decreases most rapidly, i.e., in the direction of steepest descent on the surface

$$z = \varepsilon(x, y).$$

We have for h, k small

$$\varepsilon(x+h, y+k) - \varepsilon(x, y) \doteqdot h\,\varepsilon_x + k\,\varepsilon_y$$

and the rate of descent is

$$\frac{h\,\varepsilon_x + k\,\varepsilon_y}{\sqrt{h^2+k^2}}.$$

We have to choose the ratio $h{:}k$ to maximize this. Since by the Schwarz inequality

$$|h|\,|\varepsilon_x| + |k|\,|\varepsilon_y| < \sqrt{h^2+k^2} \cdot \sqrt{\varepsilon_x^2 + \varepsilon_y^2}$$

unless (h, k) and $(\varepsilon_x, \varepsilon_y)$ are proportional, we conclude that the direction of steepest descent is given by

$$h:k=\varepsilon_x:\varepsilon_y.$$

If $(x+h, y+k)$ is on the curve then

$$h\varepsilon_x+k\varepsilon_y \doteqdot 0$$

so that the slope of the tangent is the limit

$$\lim (k/h) \doteqdot -\varepsilon_x/\varepsilon_y.$$

Thus the direction of steepest descent is that of the normal.

This argument clearly generalises to the many dimensional case.

We now return to our special problem. Here we have $\varepsilon_x=4x-6y+2$, $\varepsilon_y=-6x+10y-4$ and, at (x_0, y_0),

$$\varepsilon_x=2, \quad \varepsilon_y=-4.$$

We therefore move off from (x_0, y_0) along the line

$$x=x_0+2r, \quad y=y_0-4r. \tag{3}$$

How far along this line should we go? It is plausible that we should continue as long as ε decreases. We determine the minimum value of

$$\begin{aligned}\varepsilon(r)&=[x_0-y_0-6r]^2+[-x_0+2y_0+10r-1]^2\\&=136r^2-20r+1\\&=\left[136r^2-20r+\frac{100}{136}\right]+\frac{36}{136}\\&=[136r-10]^2/136+36/136.\end{aligned}$$

Hence the minimum value of $\varepsilon(r)$ is 9/34 occurring for $r=5/68$, i.e., for

$$x_1=-5/34, \quad y_1=10/34.$$

The geometrical interpretation of (x_1, y_1) is the following: it is the point on the line (5) which is tangent to a curve of the family (2), namely that for which the constant is 9/34.

We now repeat this process, moving from (x_1, y_1) in the direction $(4x_1-6y_1+2):(-6x_1+10y_1-4)=2:1$ to a point

$$x_2=\frac{25}{34}, \quad y_2=\frac{25}{34}$$

where $\varepsilon(x_2, y_2)=(9/34)^2$.

In the same way we find

$$x_3=805/1156, \quad y_3=940/1156, \quad \varepsilon(x_3, y_3)=(9/34)^3,$$
$$x_4=1075/1156, \quad y_4=1075/1156, \quad \varepsilon(x_4, y_4)=(9/34)^4.$$

The approach of (x_n, y_n) to $(1, 1)$ is indicated in the diagram and is linear. We note that the points (x_n, y_n) lie alternately on the lines

$$y=x, \quad 8x-13y+5=0.$$

The second line is just *below* the axis of the ellipse which has equation

$$y+1=\tfrac{1}{2}(\sqrt{5}-1)(x-1).$$

It is instructive to compare our 'discrete' approach with a 'continuous' one. We now move continuously along a curve of steepest descent. From what we have found it is clear that the differential equation to this curve or, rather, to its projection on the x, y plane, is

$$\frac{dy}{dx}=\frac{5y-3x-2}{-3y+2x+1}$$

and the initial condition is $x_0=y_0=0$. We would expect this curve to go through $(1, 1)$ and to be reasonably close to the zig-zag path already obtained.

Two approaches to this problem are available. In each we begin with a transformation $x=X+1$, $y=Y+1$ which brings the problem to the form

$$\frac{dY}{dX}=\frac{5Y-3X}{-3Y+2X}, \quad X_0=Y_0=1.$$

In the first we observe that this is homogeneous and introduce a new dependent variable V by putting $Y=VX$, $Y'=V+XV'$ which leads to a separable equation

$$\frac{dX}{X}=\frac{(2-3V)\,dV}{-3+3V+3V^2}, \quad X_0=-1, \quad V_0=1.$$

This equation can be solved by expressing the right-hand side in partial fractions. After some simplification we obtain the following result.

$$\left[\frac{Y-\alpha X}{\alpha-1}\right]^{7-3\sqrt{5}}=\left[\frac{-Y-\beta X}{\beta-1}\right]^{7+3\sqrt{5}}$$

where $\alpha=\tfrac{1}{2}(-1+\sqrt{5})$, $\beta=\tfrac{1}{2}(-1-\sqrt{5})$.

The second method is more appropriate in the present context. It is a very special case of the geometric theory of differential equations which is particularly relevant in the theory of dynamical systems. (Cf. G. Birkhoff and G. C. Rota, Ordinary differential equations, 1962.)

In this approach we interpret the diagonalization of the symmetric matrix $\mathbf{A}$ by an orthogonal similarity by a matrix $\mathbf{S}$

$$\mathbf{A}=\begin{bmatrix} 5 & -3 \\ -3 & 2 \end{bmatrix}, \quad \mathbf{S}=\begin{bmatrix} 2/\sqrt{10-2\sqrt{5}} & 2/\sqrt{10+2\sqrt{5}} \\ -2/\sqrt{10+2\sqrt{5}} & 2/\sqrt{10-2\sqrt{5}} \end{bmatrix}$$

$$\mathbf{S}^{-1}\mathbf{A}\mathbf{S}=\operatorname{diag}\left[\tfrac{1}{2}(7-3\sqrt{5}),\ \tfrac{1}{2}(7+3\sqrt{5})\right]$$

in the following way. If we write

$$\begin{bmatrix} Y \\ X \end{bmatrix}=\mathbf{S}\begin{bmatrix} \eta \\ \xi \end{bmatrix}, \qquad \begin{aligned} Y&=\eta\cos\theta+\xi\sin\theta \\ X&=-\eta\sin\theta+\xi\cos\theta \end{aligned}$$

which can be interpreted as a rotation of the axes (Y, X) to (η, ξ) through $\theta=\arctan\frac{1}{2}(\sqrt{5}-1)\doteq .55^r\doteq 32^\circ$ then the differential equation can be written as

$$\frac{d\eta}{d\xi}=\frac{(7-3\sqrt{5})\,\eta}{(7+3\sqrt{5})\,\xi}, \qquad \begin{aligned} \eta_0&=\sin\theta-\cos\theta=(\sqrt{5}-3)\big/\sqrt{10-2\sqrt{5}}\doteq -.32 \\ \xi_0&=-\sin\theta-\cos\theta=(-\sqrt{5}-1)\big/\sqrt{10-2\sqrt{5}}\doteq -1.38 \end{aligned}$$

which integrates to

$$(\eta/\eta_0)=(\xi/\xi_0)^{(7-3\sqrt{5})/(7+3\sqrt{5})}.$$

This integral is of parabolic type: instead of the usual $y^2=4ax$ or $y=\alpha x^{1/2}$ we have $\eta=\beta\xi^{.021}$.

The path of steepest descent approaches the center of the ellipse in the direction of its major axis.

We shall now discuss a somewhat similar approach to the general case. Let $\mathbf{A}$ be a positive definite matrix. We consider the solution of

$$\mathbf{A}\mathbf{x}=\mathbf{b}.$$

Let $\mathbf{r}=\mathbf{r}(\mathbf{x})$ denote the residual vector $\mathbf{b}-\mathbf{A}\mathbf{x}$. Since $\mathbf{A}$ is positive definite so is $\mathbf{A}^{-1}$ and $\mathbf{x}$ is a solution if and only if $\mathbf{r}(\mathbf{x})=0$, i.e., if and only if the quadratic form

$$f(\mathbf{x})=\mathbf{r}'\mathbf{A}^{-1}\mathbf{r} \tag{4}$$

vanishes. We calculate $f(\mathbf{x})$ as

$$\begin{aligned} f(\mathbf{x})&=(\mathbf{b}-\mathbf{A}\mathbf{x})'\mathbf{A}^{-1}(\mathbf{b}-\mathbf{A}\mathbf{x}) \\ &=\mathbf{x}'\mathbf{A}\mathbf{x}-2\mathbf{b}'\mathbf{x}+\mathbf{b}'\mathbf{A}^{-1}\mathbf{b} \end{aligned}$$

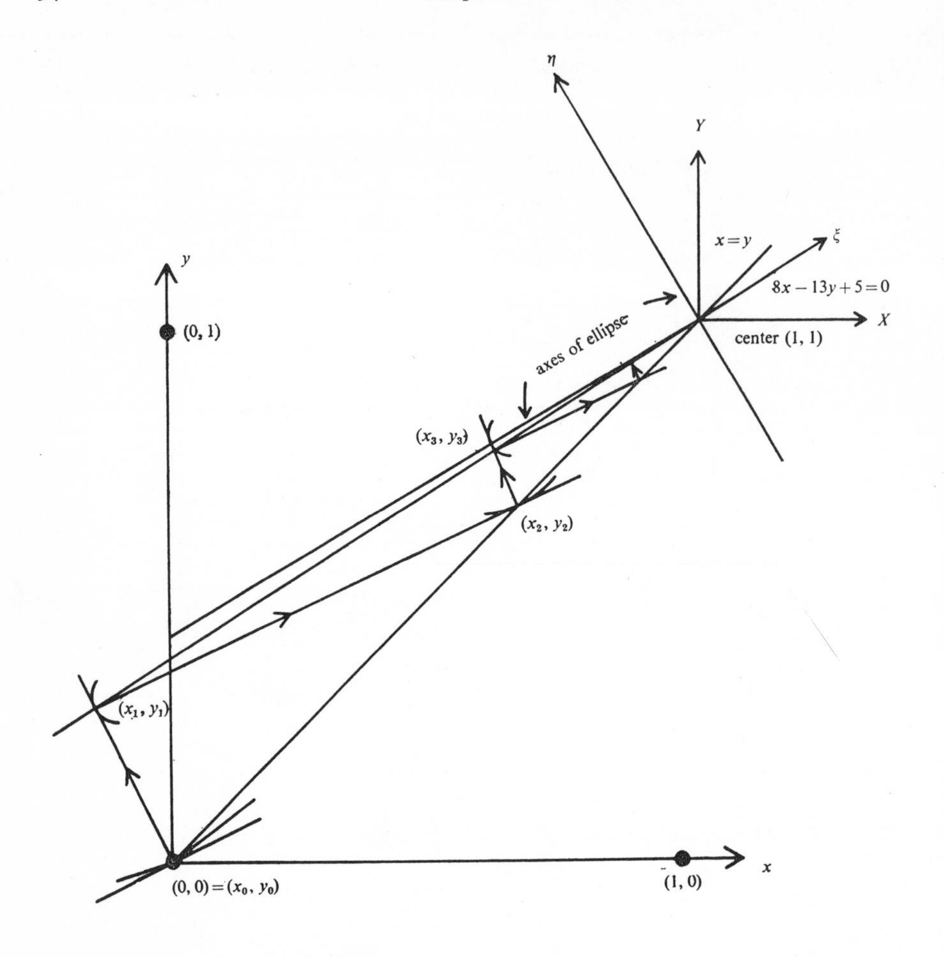

and our objective is to minimize $f(\boldsymbol{x})$ starting from an arbitrary vector $\boldsymbol{x}^{(0)}$ and proceeding along lines of steepest descents until we reach a minimum and then changing to the new line of steepest descent.

What is the direction of steepest descent at $\boldsymbol{x}^{(0)}$? As in the two-dimensional case we find that

$$\frac{\partial f}{\partial x_i} \equiv f_{x_i} = 2 r_i$$

so that the steepest direction at a point is that of the residual there. We therefore write

$$\boldsymbol{x} = \boldsymbol{x}^{(s)} + \alpha_s \boldsymbol{r}^{(s)}$$

and determine α_s by minimizing $f(\boldsymbol{x})$. This we do by "completing the square" as in the two-dimensional case. We have

$$\begin{aligned} f(\boldsymbol{x}+\alpha\boldsymbol{r}) &= (\boldsymbol{x}+\alpha\boldsymbol{r})'\boldsymbol{A}(\boldsymbol{x}+\alpha\boldsymbol{r})-2\boldsymbol{b}'(\boldsymbol{x}+\alpha\boldsymbol{r})+\boldsymbol{b}'\boldsymbol{A}^{-1}\boldsymbol{b} \\ &= \boldsymbol{x}'\boldsymbol{A}\boldsymbol{x}-2\boldsymbol{b}'\boldsymbol{x}+\boldsymbol{b}'\boldsymbol{A}^{-1}\boldsymbol{b}+\alpha^2\boldsymbol{r}'\boldsymbol{A}\boldsymbol{r}+\alpha(\boldsymbol{r}'\boldsymbol{A}\boldsymbol{x}+\boldsymbol{x}'\boldsymbol{A}\boldsymbol{r}-2\boldsymbol{b}'\boldsymbol{r}) \\ &= f(\boldsymbol{x})+\alpha^2\boldsymbol{r}'\boldsymbol{A}\boldsymbol{r}-2\alpha\boldsymbol{r}'\boldsymbol{r} \end{aligned}$$

since the scalars $\boldsymbol{r}'\boldsymbol{A}\boldsymbol{x}$, $\boldsymbol{x}'\boldsymbol{A}\boldsymbol{r}$ are equal, $\boldsymbol{A}$ being symmetrical and since $\boldsymbol{b}=$ $=\boldsymbol{r}+\boldsymbol{A}\boldsymbol{x}$ so that $-2\alpha\boldsymbol{b}'\boldsymbol{r}=-2\alpha\boldsymbol{r}'\boldsymbol{r}-2\alpha\boldsymbol{x}'\boldsymbol{A}\boldsymbol{r}$.

We now have

$$f(\boldsymbol{x}+\alpha\boldsymbol{r})=f(\boldsymbol{x})+(\boldsymbol{r}'\boldsymbol{A}\boldsymbol{r})\left[\alpha-\frac{\boldsymbol{r}'\boldsymbol{r}}{\boldsymbol{r}'\boldsymbol{A}\boldsymbol{r}}\right]^2-\frac{(\boldsymbol{r}'\boldsymbol{r})^2}{\boldsymbol{r}'\boldsymbol{A}\boldsymbol{r}}$$

which shows that the minimum value of f occurs for

$$\alpha=(\boldsymbol{r}'\boldsymbol{r})/(\boldsymbol{r}'\boldsymbol{A}\boldsymbol{r})$$

and the value here is

$$f(\boldsymbol{x})-(\boldsymbol{r}'\boldsymbol{r})^2/(\boldsymbol{r}'\boldsymbol{A}\boldsymbol{r}). \tag{5}$$

Thus we have obtained an iterative process. If $\boldsymbol{r}^{(s)}\neq 0$ then $f(\boldsymbol{x}^{(s+1)})<f(\boldsymbol{x}^{(s)})$ but this does not prove that $f(\boldsymbol{x}^{(s)})\to 0$, which implies $\boldsymbol{x}^{(s)}\to\boldsymbol{x}$. It can be proved that $f(\boldsymbol{x}^{(s)})\to 0$ linearly. The proof of this depends essentially on the Kantorovich inequality (Problem 6.19). Since, from (4), $f(\boldsymbol{x}^{(s)})=\boldsymbol{r}_s'\boldsymbol{A}^{-1}\boldsymbol{r}_s$ we have from (5)

$$\frac{f(\boldsymbol{x}^{(s+1)})}{f(\boldsymbol{x}^{(s)})}=1-\frac{(\boldsymbol{r}_s'\boldsymbol{r}_s)^2}{(\boldsymbol{r}_s'\boldsymbol{A}^{-1}\boldsymbol{r}_s)(\boldsymbol{r}_s'\boldsymbol{A}\boldsymbol{r}_s)}.$$

Now the Kantorovich inequality gives, for $\boldsymbol{x}\neq 0$,

$$1\leqq\frac{(\boldsymbol{x}'\boldsymbol{A}\boldsymbol{x})(\boldsymbol{x}'\boldsymbol{A}^{-1}\boldsymbol{x})}{(\boldsymbol{x}'\boldsymbol{x})^2}\leqq\frac{(\varkappa^{1/2}+\varkappa^{-1/2})^2}{4}$$

so that

$$\frac{4}{(\varkappa^{1/2}+\varkappa^{-1/2})^2}\leqq\frac{(\boldsymbol{x}'\boldsymbol{x})}{(\boldsymbol{x}'\boldsymbol{A}\boldsymbol{x})(\boldsymbol{x}'\boldsymbol{A}^{-1}\boldsymbol{x})}\leqq 1$$

so that

$$\frac{f(\boldsymbol{x}^{(s+1)})}{f(\boldsymbol{x}^{(s)})}\leqq 1-\frac{4}{(\varkappa^{1/2}+\varkappa^{-1/2})^2}=\frac{\varkappa+\varkappa^{-1}+2-4}{(\varkappa^{1/2}+\varkappa^{-1/2})^2}=\left[\frac{\varkappa^{1/2}-\varkappa^{-1/2}}{\varkappa^{1/2}+\varkappa^{-1/2}}\right]^2=\left(\frac{\varkappa-1}{\varkappa+1}\right)^2$$

where $\varkappa=\varkappa_2=\|\boldsymbol{A}\|_2\,\|\boldsymbol{A}^{-1}\|_2$ is the euclidean condition number of $\boldsymbol{A}$.

Thus $f(\boldsymbol{x}^{(s)})\to 0$, at worst geometrically with common ratio $((\varkappa-1)/(\varkappa+1))^2$. The larger $\varkappa$, i.e., the worse the condition of $\boldsymbol{A}$ is, the slower is the convergence of the steepest descent process — once again validating the condition number as a figure of demerit for a matrix.

4. Iterative improvement of approximate inverses

It is well-known that the sequence defined for an arbitrary x_0, such that $0<x_0<2N^{-1}$, by

$$x_{n+1}=x_n(2-Nx_n), \quad n=0, 1, 2, \ldots,$$

where $0<N\leqq 1$, converges rapidly to N^{-1}. In fact, if $\varepsilon_n=x_n-N^{-1}$, we find that $\varepsilon_{n+1}=-N\varepsilon_n^2$. This recurrence relation comes from the Newton formula $x_{n+1}=x_n-(f(x_n)/f'(x_n))$ applied when $f(x)=N-x^{-1}$.

This idea can be transferred essentially letter by letter into a scheme for improving an approximate inverse $\mathbf{X}_0$ of a non-singular matrix $\mathbf{A}$. We discuss this in detail. The general stage can be better written in the form

$$\mathbf{X}_{n+1}=\mathbf{X}_n+\mathbf{X}_n(\mathbf{I}-\mathbf{A}\mathbf{X}_n) \tag{6}$$

(where the second term on the right is the correction term) and, from the point of view of theoretical arithmetic, this gives

$$\begin{aligned}\mathbf{I}-\mathbf{A}\mathbf{X}_{n+1}&=\mathbf{I}-\mathbf{A}\mathbf{X}_n-\mathbf{A}\mathbf{X}_n(\mathbf{I}-\mathbf{A}\mathbf{X}_n)\\&=(\mathbf{I}-\mathbf{A}\mathbf{X}_n)^2.\end{aligned}$$

This means that if we have a sufficiently good first approximation, say,

$$\|\mathbf{I}-\mathbf{A}\mathbf{X}_0\|=\varepsilon<1,$$

then

$$\|\mathbf{I}-\mathbf{A}\mathbf{X}_n\|\leqq\varepsilon^{2^n},$$

so that we have quadratic convergence. In practical computation we get very satisfactory results provided the correction term in (6) is computed with care.

If our attempt to solve the system

$$\mathbf{A}\mathbf{x}=\mathbf{b}$$

produces an alleged or machine solution $\bar{\mathbf{x}}$ with residual vector

$$\mathbf{r}=\mathbf{b}-\mathbf{A}\bar{\mathbf{x}}$$

we can obtain a correction ξ

$$\mathbf{x}=\bar{\mathbf{x}}+\xi$$

to the machine solution by the solution of

$$\mathbf{A}\xi=\mathbf{r}.$$

If the solution now obtained, $\mathbf{x}+\bar{\xi}$, is not satisfactory we can repeat this process until an acceptable solution is obtained.

Several remarks are appropriate here. First, if we are solving the systems by a triangular decomposition of $\mathbf{A}$, the decomposition has only to be done once — and the actual additional expense per refinement is only $\mathcal{O}(n^2)$ operations. The second remark is that it is imperative to compute the residuals by accumulating the partial sums to the residuals by accumulating the partial sums to double precision, and then rounding off to single precision.

Since the facility of double precision accumulation is not generally readily available, i.e., it has to be programmed, we do not insist on a worked example. However those interested can experiment, using ill-conditioned systems in which the exact solution is known.

Chapter 9, Problems

9.1. Find the general residual vectors $\boldsymbol{\varepsilon}^{(r)}$ explicitly in the case of the solution of the system

$$\begin{bmatrix} 1 & -\frac{1}{2} \\ -\frac{1}{2} & 1 \end{bmatrix} \boldsymbol{x} = \begin{bmatrix} \frac{1}{2} \\ \frac{1}{2} \end{bmatrix}$$

by the Jacobi process and by the Gauss—Seidel process.

9.2. (Collatz.) Find the inverse of the matrix

$$\begin{bmatrix} 1 & 0 & 0 \\ -a & 1 & 0 \\ -b & -c & 1 \end{bmatrix}.$$

Discuss the convergence of the Jacobi and Gauss—Seidel processes for the solution of $\boldsymbol{Ax}=\boldsymbol{b}$ when

$$\boldsymbol{A}=\begin{bmatrix} 1 & -2 & 2 \\ -1 & 1 & -1 \\ -2 & -2 & 1 \end{bmatrix} \quad \text{and when} \quad \boldsymbol{A}=\begin{bmatrix} 1 & \frac{1}{2} & -\frac{1}{2} \\ -1 & 1 & -1 \\ \frac{1}{2} & \frac{1}{2} & 1 \end{bmatrix}.$$

9.3. Draw a rough graph of $\varrho_\omega=\varrho(\mathscr{L}_\omega)$, where $\mathscr{L}_\omega$ is defined on p. 89.

9.4. The matrix $\boldsymbol{X}=\boldsymbol{X}_0=\begin{bmatrix} -3.9 & 4.1 & -0.9 \\ 4.1 & -5.1 & 1.9 \\ -0.9 & 1.9 & -1.1 \end{bmatrix}$ is an approximate inverse of $\boldsymbol{A}=\begin{bmatrix} 1 & 2 & 3 \\ 2 & 3 & 4 \\ 3 & 4 & 4 \end{bmatrix}$; the exact inverse $\boldsymbol{A}^{-1}=\begin{bmatrix} -4 & 4 & -1 \\ 4 & -5 & 2 \\ -1 & 2 & -1 \end{bmatrix}$.

Calculate $\boldsymbol{X}_1, \boldsymbol{X}_2, \boldsymbol{X}_3$, where $\boldsymbol{X}_{n+1}=\boldsymbol{X}_n(2\boldsymbol{I}-\boldsymbol{A}\boldsymbol{X}_n)$, $n>0$ and also calculate, for $n=0, 1, 2, 3$

$$\|\boldsymbol{I}-\boldsymbol{A}\boldsymbol{X}_n\|$$

and

$$\|A^{-1} - X_n\|.$$

9.5. Write a program for carrying out the iterative correction process and apply it, e.g., in the case of H_n, the Hilbert matrix or W, the Wilson matrix.

9.6. Derive the characteristic values given for the Young matrices, corresponding to the special 4×4 matrix A of p. 87, either (a) by use of Problem 4.10 or (b), by use of Williamson's Theorem, Problem 6.16.

9.7. Show that both the Jacobi and Gauss—Seidel methods are convergent when the matrix has a strictly dominant diagonal.

9.8. Show that the Gauss—Seidel method is convergent when the matrix is positive definite.

9.9. Apply the Gauss—Seidel method to the first system of Problem 6.3.

9.10. Fill in the following table from the Fitzgerald papers.

(K. E. Fitzgerald, *Error estimates for the solution of linear algebraic systems,* J. Research Nat. Bur. Standards *74B*, 251—310, (1970), and *Comparison of some FORTRAN programs for matrix inversion,* ibid. *78B,* 15—33, (1974).)

Matrix	Condition Number	$\|I-AX\|_F$	$\|I-XA\|_F$	$\|I-A\mathcal{X}\|_F$	$\|I-\mathcal{X}A\|_F$
T_{20}^3					
T_{100}					
T_{100}^2					
H_6					
H_8					
H_{10}					

Comment on the results.

(Note that Fitzgerald's T is our A_8 and his H is our A_{13}; also $\mathcal{X}$ indicates an iterative refinement of the alleged inverse X.)

9.11. Exhibit the three iterative methods (of Jacobi, Gauss—Seidel and Young) as special cases of the following general scheme:

If A is non-singular and has $a_{ii}\neq 0$, $i=1, 2, \ldots, n$ and if $A=M-N$ where M is non-singular, consider the sequence of vectors $x^{(r)}$ generated from an initial guess $x^{(0)}$ by

$$x^{(r+1)} = M^{-1}Nx^{(r)} + M^{-1}b, \quad r \geqq 1.$$

CHAPTER 10

Application: Solution of a Boundary Value Problem

The two-point boundary value problem

$$-y'' - \lambda y = 0 \tag{1}$$
$$y(0) = y(1) = 0$$

occurs, e.g. when we separate the variables in the wave equation

$$\frac{\partial^2 z}{\partial x^2} = \frac{\partial^2 z}{\partial t^2}$$

by assuming that

$$z(x, t) = y(x) \exp i \sqrt{\lambda}\, t.$$

If the boundary conditions are

$$z(0, t) = z(1, t) = 0,$$
$$z(x, 0) \text{ given}$$

the problem can be interpreted in terms of the vibrations of a uniform string with fixed end-points ($x=0$, $x=1$) and with initial displacement $z(x, 0)$.

The problem (1) has clearly non-trivial solutions if and only if λ is a "characteristic value" of the problem, in this case,

$$\lambda = (r\pi)^2, \quad r = 1, 2, \ldots. \tag{2}$$

The corresponding characteristic functions are

$$y(x) = \sin r\pi x.$$

The general problem of which this is the prototype is called a Sturm—Liouville problem for it was studied by J. C. F. Sturm (1803—1855) and J. Liouville (1809—1882). This problem has already been mentioned briefly in Part I of this book; we shall now discuss it further, by two different methods. For a modern account of the theory see, e.g., G. Birkhoff, G. C. Rota, *Ordinary Differential Equations,* 1962, or H. B. Keller, *Numerical Methods for Two-point Boundary-Value Problems,* 1968.

The general theory guarantees (under certain smoothness conditions) that the problem

$$\left.\begin{aligned} -y'' - \lambda q(x) y &= 0 \\ y(0) = y(1) &= 0 \end{aligned}\right\} \tag{3}$$

where $q(x)>0$ in $(0, 1)$, has an infinite sequence of characteristic values λ_n

$$0<\lambda_1\leqq\lambda_2\leqq\lambda_3\leqq\cdots$$

which tend to infinity and, corresponding to each, there is a characteristic function $y_n(x)$ such that

$$\left.\begin{aligned}-y_n''-\lambda_n q(x)\,y_n=0\\ y_n(0)=y_n(1)=0\end{aligned}\right\};$$

further $y_n(x)$ has exactly $n-1$ zeros in $(0, 1)$. Also, the solutions have an orthogonality property.

We consider "discretizing" the problem and discuss the error made in discretizing: in general we may expect only a finite set of characteristic values Λ_n of the discrete problem, and we discuss how are these related to the λ's and similarly for the characteristic functions.

1. Method I

We shall discuss only the prototype problem (1), and in this case we can find the Λ_n exactly.

A natural discretization of (1) is the following: find a non-zero vector $y=[Y_0=0, Y_1, \dots, Y_n, Y_{n+1}=0]'$ such that

$$-\frac{Y_{r+1}-2Y_r+Y_{r-1}}{h^2}-\Lambda Y_r=0,\quad r=1, 2, \dots, n.$$

These n equations can be combined into a matrix equation

$$\begin{bmatrix} 2 & -1 & & & & \\ -2 & 2 & -1 & & & \\ & & \cdots & & & \\ & & & -1 & 2 & -1 \\ & & & & -1 & 2 \end{bmatrix}\hat{y}=\Lambda h^2\,\hat{y}$$

where now

$$\hat{y}=[Y_1, Y_2, \dots, Y_{n-1}, Y_n]'.$$

Using the results of Problem 5.13. (iv) we see that there are solutions provided

$$\Lambda_r=4h^{-2}\sin^2\frac{r\pi}{2(n+1)},\quad r=1, 2, \dots, n$$

$$=4(n+1)^2\sin^2\frac{r\pi}{2(n+1)}.$$

For $r \ll n$, since $\sin x \doteqdot x$ for x near 0, we have

$$\Lambda_r \doteqdot r^2 \pi^2$$

in agreement with (2). Estimates of the error can be found, and we observe that the agreement deteriorates as r increases.

The results of Problem 5.13. (iv) also enable us to compare the characteristic functions and vectors and we observe a similar deterioration as r increases.

If $q(x)$ is not constant the analysis would be more difficult to carry through — it is however fairly easy to show theoretically that there will always be n real characteristic values.

2. Method II

We shall now discuss a different approach to this problem. The method of Galerkin (1871—1945) has proved to be of considerable value in many practical problems, particularly when combined with recent developments in the theory of approximation.

We seek an approximation to the solution of (3) in the form

$$\omega(x) = \omega_{\beta}(x) = \sum \beta_i b_i(x)$$

where the $b_i(x)$, $i = 1, 2, \dots, n$ are linearly independent functions, which we shall suppose to satisfy $b_i(0) = b_i(1) = 0$, $i = 1, 2, \dots, n$. Since our problem is homogeneous we may suppose $\|\boldsymbol{\beta}\|_2 = 1$.

It is clear that, in general, we cannot choose the vector $\boldsymbol{\beta}$ and the constant λ to ensure that

$$-\omega''(x) - \lambda q(x)\omega(x) = 0.$$

We can, however, ask that the residual of $\omega(x)$ be orthogonal to all the basis functions $b_j(x)$, i.e.,

$$\int_0^1 [\omega''(x) + \lambda q(x)\omega(x)] b_i(x)\, dx = 0, \quad i = 1, 2, \dots, n.$$

If we write $\mathbf{A} = [a_{ij}]$, $\mathbf{B} = [b_{ij}]$ where

$$a_{ij} = \int_0^1 b_i'(x) b_j'(x)\, dx, \quad b_{ij} = \int_0^1 q(x) b_i(x) b_j(x)\, dx$$

these conditions can be restated in the following way, after an integration by parts:

$$-\sum_j \beta_j a_{ij} + \lambda \sum_j \beta_j b_{ij} = 0, \quad i = 1, 2, \dots, n.$$

Thus we have to determine a vector $\boldsymbol{\beta}$ with $\|\boldsymbol{\beta}\|_2 = 1$ and a constant λ such that

$$\mathbf{A}\boldsymbol{\beta} = \lambda \mathbf{B}\boldsymbol{\beta}.$$

This is the so-called general characteristic value problem, reducing to the usual one in case $\mathbf{B}=\mathbf{I}$.

It will surely be advantageous in the solution of the problem if the matrices $\mathbf{A}$, $\mathbf{B}$ are sparse and it is natural to try to choose a basis which ensures this. If we choose the $b_r(x)$, $r=1, 2, \ldots, n$ to be the functions indicated:

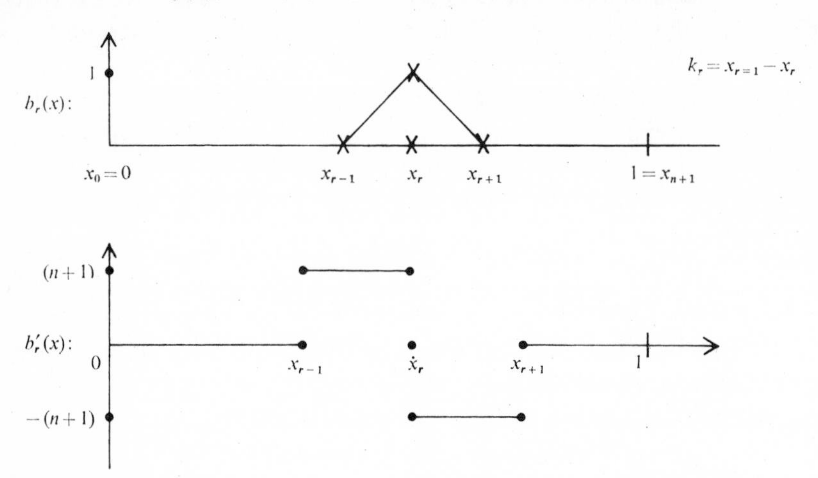

both the matrices $\mathbf{A}$, $\mathbf{B}$ will clearly be tridiagonal no matter what $q(x)$ is. If we choose $q(x)\equiv 1$ we find

$$a_{r,r}=\int_0^1 (b_r'(x))^2\,dx=(n+1)^2\int_{x_{r-1}}^{x_{r+1}} 1\,dx=2(n+1);$$

$$a_{r,r+1}=\int_0^1 b_r'(x)b_{r+1}'(x)\,dx=(n+1)^2\int_{x_r}^{x_{r+1}}(-1)(1)\,dx=-(n+1);$$

$$\begin{aligned}b_{r,r}=\int_0^1 (b_r(x))^2\,dx&=2\int_{x_r}^{x_{r+1}}[1-(n+1)(x-x_r)]^2\,dx=2\int_0^{h_r}(1-(n+1)t)^2\,dt\\&=2[t-(n+1)t^2+(n+1)^2(t^3/3)]_0^{1/(n+1)}\\&=2\left[\frac{1}{n+1}-\frac{1}{n+1}+\frac{1}{3(n+1)}\right]\\&=\frac{2}{3}\cdot\frac{1}{n+1};\end{aligned}$$

$$\begin{aligned}b_{r,r+1}=\int_0^1 b_r(x)b_{r+1}(x)\,dx&=\int_{x_r}^{x_{r+1}}[1-(n+1)(x-x_r)][(n+1)(x-x_r)]\,dx\\&=\frac{1}{6}\cdot\frac{1}{n+1}.\end{aligned}$$

We have

$$\mathbf{A}=(n+1)\begin{bmatrix} 2 & -1 & & & \\ -1 & 2 & -1 & & \\ & & \ddots & & \\ & & -1 & 2 & -1 \\ & & & -1 & 2 \end{bmatrix}, \quad 6(n+1)\mathbf{B}=\begin{bmatrix} 4 & 1 & & & \\ 1 & 4 & 1 & & \\ & & \ddots & & \\ & & 1 & 4 & 1 \\ & & & 1 & 4 \end{bmatrix}.$$

so that, where $\mathbf{C}_n$ is the $n \times n$ second difference matrix $\mathbf{A}_8$,

$$\mathbf{A}=(n+1)\mathbf{C}_n, \quad \mathbf{B}=[6I-\mathbf{C}_n]/[6(n+1)], \quad \mathbf{B}^{-1}\mathbf{A}=(n+1)^2\mathbf{C}_n/[1-(\mathbf{C}_n/6)].$$

Since $\mathbf{B}$ is non-singular the problem $\mathbf{A}\boldsymbol{\beta}=\lambda\mathbf{B}\boldsymbol{\beta}$ can be replaced by

$$\mathbf{B}^{-1}\mathbf{A}\boldsymbol{\beta}=\lambda\boldsymbol{\beta},$$

i.e., the ordinary characteristic value problem for $\mathbf{B}^{-1}\mathbf{A}$. Since both $\mathbf{B}$, $\mathbf{A}$ are polynomials in $\mathbf{C}_n$ the characteristic values are just

$$\lambda=\lambda_k\equiv\frac{(n+1)^2\gamma_k}{1-\frac{1}{6}\gamma_k}$$

where

$$\gamma_k=2(1-\cos 2k\theta), \quad k=1,2,\dots,n, \ \theta=\pi/2(n+1),$$

are the characteristic values of $\mathbf{C}_n$. (See Problem 5.13. (iv).)

These, therefore, are the approximate characteristic values of (1). For $r \ll n$ we have

$$\gamma_r=4\sin^2 r\theta \doteqdot r^2\pi^2/n^2$$
$$\lambda_r \doteqdot r^2\pi^2$$

in agreement with (2). The approximations to the characteristic *functions* can be obtained from the results of Problem 5.13. (iv): they are piecewise linear functions whose values at the points $k/(n+1)$, $k=0, 1, \dots, n+1$ are

$$0, \beta_1^{(r)}, \beta_2^{(r)}, \dots, \beta_{n-1}^{(r)}, \beta_n^{(r)}, 0$$

where $\boldsymbol{\beta}^{(r)}$ is now a characteristic vector of $\mathbf{C}_n$, i.e.,

$$\boldsymbol{\beta}^{(r)}=[\sin 2r\theta, \sin 4r\theta, \dots, \sin 2nr\theta]'.$$

3. Method III

See Problem 10.1. (Cf. B. Wendroff, *Theoretical Numerical Analysis*, Academic Press, 1966.)

Chapter 10, Problem

10.1. Obtain an approximate solution to the problem (3) by the following Rayleigh—Ritz process. It is well known from the theory of differential equations that the fundamental solution to (3) is given by the function $u_1(x)$ which minimizes

$$R(u)=\frac{\int_0^1 (u'(x))^2\,dx}{\int_0^1 q(x)(u(x))^2\,dx}$$

over all differentiable functions $u(x)$ which satisfy the boundary conditions. To get an approximation to u, by an algebraic process, take

$$U(x)=U_c(x)=\sum c_i b_i(x)$$

where the $b_i(x)$ satisfy the boundary conditions, and take the minimum over all vectors $c\in R_n$: that is, determine

$$\min \frac{\sum\sum A_{ij}c_ic_j}{\sum\sum B_{ij}c_ic_j}$$

where

$$\begin{aligned} A_{ij}&=\int_0^1 b_i'(x)\,b_j'(x)\,dx, \\ & \qquad\qquad\qquad\qquad i,j=1,2,\dots,n, \\ B_{ij}&=\int_0^1 q(x)\,b_i(x)\,b_j(x)\,dx. \end{aligned}$$

Take the $b_i(x)$ to be the piecewise linear functions used in the discussion of the Galerkin process.

CHAPTER 11

Application: Least Squares Curve Fitting

We want to indicate some of the arithmetic dangers of this process. However we begin by noting that serious consideration must always be given as to whether this is the appropriate problem: e.g., perhaps exponential fitting with a maximum norm is more suitable than polynomial fitting with a euclidean norm.

The problem is to find that polynomial of degree k

$$y=y_c(x)=c_0+c_1x+\ldots+c_kx^k$$

such that for an assigned set of *distinct* abscissas

$$x_1, \ldots, x_m$$

and values

$$f_1, \ldots, f_m$$

the discrepancy

$$\sum_{i=1}^{m} |y(x_i)-f_i|^2$$

is least. This is a new problem only if $m \geqq k+1$ and we assume that this is the case.

We use the following notation:

$$\boldsymbol{f}=\begin{bmatrix} f_1 \\ f_2 \\ \vdots \\ f_m \end{bmatrix}, \quad \boldsymbol{c}=\begin{bmatrix} c_0 \\ c_1 \\ \vdots \\ c_k \end{bmatrix}, \quad \mathbf{Q}=[\boldsymbol{q}^{(0)}, \boldsymbol{q}^{(1)}, \ldots, \boldsymbol{q}^{(k)}], \; \boldsymbol{q}^{(i)}=\begin{bmatrix} x_1^i \\ x_2^i \\ \vdots \\ x_m^i \end{bmatrix}.$$

Our problem is clearly to minimize $\|\boldsymbol{f}-\mathbf{Q}\boldsymbol{c}\|_2^2$ over $\boldsymbol{c}$. The solution to this problem is $\boldsymbol{c}=(Q'Q)^{-1}Q'f$. To see this we proceed as follows. We have

$$\|\boldsymbol{f}-\mathbf{Q}\boldsymbol{c}\|_2^2=(\boldsymbol{f}-\mathbf{Q}\boldsymbol{c})'(\boldsymbol{f}-\mathbf{Q}\boldsymbol{c})=\boldsymbol{f}'\boldsymbol{f}-2\boldsymbol{c}'(\mathbf{Q}'\boldsymbol{f})+\boldsymbol{c}'\mathbf{Q}'\mathbf{Q}\boldsymbol{c}.$$

We now follow an appropriate generalization of the "completing the square" technique used in elementary algebra: since

$$q(x)=1+2x+3x^2 \equiv (\sqrt{3}x+(1/\sqrt{3}))^2+(2/3)$$

it is clear that $q(x)$ has a minimum (2/3) attained if and only if $x=-(1/3)$. In the present case we readily verify that

$$\|\boldsymbol{f}-\mathbf{Q}\boldsymbol{c}\|_2^2=$$
$$=[\boldsymbol{c}-(\mathbf{Q}'\mathbf{Q})^{-1}(\mathbf{Q}'\boldsymbol{f})]'\mathbf{Q}'\mathbf{Q}[\boldsymbol{c}-(\mathbf{Q}'\mathbf{Q})^{-1}(\mathbf{Q}'\boldsymbol{f})]+\boldsymbol{f}'\boldsymbol{f}-(\mathbf{Q}'\boldsymbol{f})'(\mathbf{Q}'\mathbf{Q})^{-1}(\mathbf{Q}'\boldsymbol{f})$$

and on the right-hand side the first term is the only one which depends on $\boldsymbol{c}$ and, being of the form $\boldsymbol{v}'\mathbf{Q}'\mathbf{Q}\boldsymbol{v}$ is non-negative and zero if and only if $\boldsymbol{v}=0$.

Indeed the matrix $\mathbf{Q}'\mathbf{Q}$ which is obviously symmetric is also positive definite. For the corresponding quadratic form $\boldsymbol{v}'(\mathbf{Q}'\mathbf{Q})\boldsymbol{v}=(\mathbf{Q}\boldsymbol{v})'(\mathbf{Q}\boldsymbol{v})=\|\mathbf{Q}\boldsymbol{v}\|_2^2>0$ unless $\mathbf{Q}\boldsymbol{v}=0$ and this can only happen if $\boldsymbol{v}=0$ because $\mathbf{Q}\boldsymbol{v}=0$ means $v_0+v_1x+\ldots+v_kx^k=0$ for $x=x_1, x_2, \ldots, x_m$ and a polynomial of degree k which vanishes at $m>k$ distinct points is necessarily identically zero. Hence $\mathbf{Q}'\mathbf{Q}$ is positive definite and so non-singular.

Hence $\|\boldsymbol{f}-\mathbf{Q}\boldsymbol{c}\|$ is minimized when

$$\boldsymbol{c}=(\mathbf{Q}'\mathbf{Q})^{-1}(\mathbf{Q}'\boldsymbol{f}),$$

i.e. when

$$(\mathbf{Q}'\mathbf{Q})\boldsymbol{c}=\mathbf{Q}'\boldsymbol{f},$$

i.e. when $\boldsymbol{c}$ is a solution of the "normal equations".

We can also obtain the result by calculus methods. We want the minimum of

$$E=\boldsymbol{f}'\boldsymbol{f}-2\boldsymbol{c}'(\mathbf{Q}'\boldsymbol{f})+\boldsymbol{c}'\mathbf{Q}'\mathbf{Q}\boldsymbol{c}$$

regarded as a function of the $k+1$ parameters $c_0, c_1, \ldots, c_k$. We therefore equate the partial derivatives of E to zero. The first term of E is a constant, the second a scalar product and the third a quadratic form. We have, for $i=0, 1, \ldots, k$,

$$\frac{\partial E}{\partial c_i}\equiv -2(Q'f)_i+2(Q'Q)_{ii}c_i+2\sum_j{}'(Q'Q)_{ij}c_j=0.$$

These scalar equations can be combined into

$$(\mathbf{Q}'\mathbf{Q})\boldsymbol{c}=\mathbf{Q}'\boldsymbol{f},$$

the normal equations. If we differentiate again we get

$$\frac{\partial^2 E}{\partial c_i\,\partial c_j}=2(Q'Q)_{ij}$$

and the positive definiteness of $\mathbf{Q}'\mathbf{Q}$ is the standard sufficient condition for a minimum.

Let us consider the element $[\mathbf{Q}'\mathbf{Q}]_{ij}$. This is

$$\sum_{l=1}^{m} [\mathbf{Q}']_{il}[\mathbf{Q}]_{lj}$$

$$= \sum_{l=1}^{m} [\mathbf{Q}]_{li}[\mathbf{Q}]_{lj}$$

$$= \sum_{l=1}^{m} x_l^i x_l^j = \sum_{l=1}^{m} x_l^{i+j}.$$

Now if the x_l are chosen fairly uniformly in the interval [0, 1] we have

$$m^{-1} \sum x_l^{i+j} \doteqdot \int_0^1 x^{i+j}\, dx = \frac{1}{i+j+1}.$$

Thus the matrix to be inverted to solve for $\mathbf{c}$ is the Hilbert matrix, the condition of which we know to be bad. This argument indicates the delicacy of our problem. Alternative approaches have been given by G. E. Forsythe and standard algorithms have been prepared, e.g., by Businger and Golub.

REMARK: We encounter the same problem if we start off with an *over-determined system of linear equations,* say $m>k$ equations in k unknowns, which we can write as

$$\mathbf{Q}\mathbf{c} \stackrel{?}{=} \mathbf{f}.$$

We look for the solution $\mathbf{c}$ which minimizes the residual

$$\mathbf{r} = \mathbf{Q}\mathbf{c} - \mathbf{f}$$

in the euclidean norm. This is the problem already discussed except that now the $m(k+1)$ elements of $\mathbf{Q}$ are *arbitrary.*

The solution of the normal equations

$$\mathbf{Q}'\mathbf{Q}\mathbf{c} = \mathbf{Q}'\mathbf{f} \tag{1}$$

can be carried out by the *Cholesky method.* We factorize the positive definite symmetric matrix $\mathbf{Q}'\mathbf{Q}$ as

$$\mathbf{Q}'\mathbf{Q} = \mathbf{L}\mathbf{L}' \tag{2}$$

and then solve

$$\mathbf{L}\mathbf{y} = \mathbf{Q}'\mathbf{f},$$

$$\mathbf{L}'\mathbf{c} = \mathbf{y}.$$

A mathematically equivalent (but sometimes numerically more desirable) method of solving these equations is by *orthogonalisation.* If we express $\mathbf{Q}$

in the form (cf. Problems 4.6 and 8.14)

$$\mathbf{Q}=\boldsymbol{\Phi}\mathbf{U}$$

where $\boldsymbol{\Phi}$ is orthogonal and $\mathbf{U}$ upper triangular we obtain, on symmetrization,

$$\begin{aligned}\mathbf{Q}'\mathbf{Q}&=(\boldsymbol{\Phi}\mathbf{U})'\boldsymbol{\Phi}\mathbf{U}\\ &=\mathbf{U}'\boldsymbol{\Phi}'\boldsymbol{\Phi}\mathbf{U} \qquad (3)\\ &=\mathbf{U}'\mathbf{U}.\end{aligned}$$

However, under standard normalization, the Cholesky factorization is unique and comparison of (2) and (3) shows that $\mathbf{L}=\mathbf{U}'$. We complete this version of the solution by finding first

$$\boldsymbol{y}=\boldsymbol{\Phi}'\boldsymbol{f}$$

and then solving for $\boldsymbol{c}$

$$\mathbf{U}\boldsymbol{c}=\boldsymbol{y}.$$

The last two equations give $\mathbf{U}\boldsymbol{c}=\boldsymbol{\Phi}'\boldsymbol{f}$ and premultiplication by $\mathbf{U}'$ gives $\mathbf{U}'\mathbf{U}\mathbf{C}=\mathbf{U}'\boldsymbol{\Phi}'\boldsymbol{f}$ which are the normal equations.

For an example see Problem 11.1.

Chapter 11, Problems

11.1. Find the best linear fit in the least squares sense to the data

$$x_1=1 \quad f_1=1$$
$$x_2=2 \quad f_2=3$$
$$x_3=3 \quad f_3=1$$

Use both the direct solution of the normal equations and the orthogonalization method.

11.2, Test the fitting program in the computer library in the following way. Take sets of six random numbers, $p_0, p_1, \ldots, p_5$ say between 0 and 1, generated by the computer. Evaluate, for $n=0(1)20$

$$V_n=p_0+p_1 n+p_2 n^2+\ldots+p_5 n^5.$$

Use the set $x_r=r, f_r=V_r$, $r=0, 1, \ldots, 20$ as input data for the program. The output should be, were it not for rounding errors in the evaluation of the V's and in the program itself, just

$$c_0=p_0,\ c_1=p_1,\ \ldots,\ c_5=p_5.$$

Is it?

11.3. Repeat Problem 11.2 in the case

(1) $p_0=p_1=p_2=p_3=p_4=p_5=1$

(2) $p_0=1, \quad p_1=10^{-1}, \quad p_2=10^{-2}, \quad p_3=10^{-3}, \quad p_4=10^{-4}, \quad p_5=10^{-5}$.

11.4. Find the best linear fit to the data of Problem 11.1 if best is interpreted in the Chebyshev sense:

$$\min_{a,\,b} \max_{i=1,\,2,\,3} |a x_i + b - f_i|.$$

11.5. Find the best fit, in the least squares sense, to the data

$$\begin{array}{lllllllllll} x_i: & 0 & 1 & 2 & 3 & 4 & 5 & 6 & 7 & 8 & 9, \\ f_i: & 0 & 2 & 2 & 5 & 5 & 6 & 7 & 7 & 7 & 10, \end{array}$$

by a polynomial of degree at most 6.

11.6. Show that the solution to a least squares problem is not necessarily unique by discussing the system

$$x+y=1, \quad 2x+2y=0 \quad -x-y=2.$$

11.7. Discuss the least squares solution of the problem

$$x+y=1, \quad 2x=0, \quad -x+3y=2.$$

CHAPTER 12

Singular Value Decomposition and Pseudo-Inverses

The singular value decomposition was probably first introduced by J. W. Gibbs and pseudo-inverses probably by E. H. Moore. The importance of these concepts in numerical algebra (in particular in statistical applications) has been emphasized notably by I. J. Good and by G. H. Golub. There are many ways of defining these objects, and we choose one of those appropriate in the present context.

Theorem 12.1. *Let* $\mathbf{A}$ *be an* $m \times n$ *matrix of complex elements. Suppose* $\mathbf{A}$ *has rank* r *and that* $m \leqq n$. *Then there exist unitary matrices* $\mathbf{U}, \mathbf{V}$ *such that*

$$\mathbf{U}^* \mathbf{A} \mathbf{V} = \Sigma \quad \text{or} \quad \mathbf{A} = \mathbf{U} \Sigma \mathbf{V}^* \tag{1}$$

where the $m \times n$ *matrix* Σ *has as its first* r *diagonal elements* $\sigma_1, \ldots, \sigma_r$ *the positive singular values of* $\mathbf{A}$, *and has all other elements zero.*

We recall that the *singular values* of $\mathbf{A}$ are the non-negative square roots of the characteristic values of the non-negative definite hermitian matrix $\mathbf{A}\mathbf{A}^*$. The representation (1) is called the *singular value decomposition* of $\mathbf{A}$; there are several variants of this definition in use. It is sometimes convenient to introduce a *pair of singular vectors,* $\boldsymbol{u}$ of dimension n, $\boldsymbol{v}$ of dimension m corresponding to a singular value μ of $\mathbf{A}$; these are defined by

$$\mathbf{A} \boldsymbol{u} = \mu \boldsymbol{v}$$

$$\mathbf{A}^* \boldsymbol{v} = \mu \boldsymbol{u}.$$

Proof. We shall assume $m \leqq n$. Then [compare Problem 12.1] $\mathbf{A}\mathbf{A}^*$ has rank $r \leqq m \leqq n$. In our diagrams below we have assumed $r < m < n$ — the appropriate changes when there is one or more equality are clear.

We can find a unitary $m \times m$ matrix $\mathbf{U}$ which diagonalizes $\mathbf{A}\mathbf{A}^*$

$$\mathbf{U}^* (\mathbf{A}\mathbf{A}^*) \mathbf{U} = \operatorname{diag}[\sigma_1^2, \ldots, \sigma_r^2, 0, \ldots, 0]. \tag{2}$$

We write (2) as

$$\mathbf{F}\mathbf{F}^* = \begin{bmatrix} \mathbf{D}_1^2 & 0 \\ 0 & 0 \end{bmatrix} \tag{3}$$

where $\mathbf{F}=\mathbf{U}^*\mathbf{A}$ and $\mathbf{D}_1=\text{diag}(\sigma_1,\ldots,\sigma_r)$. From (3) it is clear that the first r rows $f_1^*, f_2^*, \ldots, f_r^*$ of $\mathbf{F}$ are orthogonal and the remaining rows are zero vectors. We can normalize these by writing

$$v_i=\sigma_i^{-1}f_i, \quad i=1,2,\ldots,r.$$

The vectors $v_1, v_2,\ldots, v_r$ are independent and orthonormal. We can find vectors $v_{r+1}, v_{r+2}, \ldots, v_n$ which with $v_1, v_2, \ldots, v_r$ form an independent set and, moreover, in virtue of the Gram—Schmidt process, we may assume the whole set orthonormal.

Now write

$$\mathbf{V}=[v_1, v_2, \ldots, v_r | v_{r+1}, v_{r+2}, \ldots, v_n]$$
$$=[\mathbf{V}_1, \mathbf{V}_2].$$

Observe that $\mathbf{V}_1^*=\mathbf{D}^{-1}\mathbf{F}_1$, where $\mathbf{F}_1$ consists of the first r rows of $\mathbf{F}$ and $\mathbf{F}_2$ the remaining $m-r$.

It is clear that

$$\mathbf{V}^*\mathbf{V}=\mathbf{I}$$

and that $\mathbf{V}_1^*\mathbf{V}_1=\mathbf{I}_r$, $\mathbf{V}_1^*\mathbf{V}_2=0$, $\mathbf{V}_2^*\mathbf{V}_1=0$, $\mathbf{V}_2^*\mathbf{V}_2=\mathbf{I}_{n-r}$.

The product $\mathbf{U}^*\mathbf{A}\mathbf{V}$ can be written as

$$\begin{bmatrix}\mathbf{F}_1\\0\end{bmatrix}[\mathbf{F}_1^*\mathbf{D}_1^{-1},\ \mathbf{V}_2]=\begin{bmatrix}\mathbf{F}_1\mathbf{F}_1^*\mathbf{D}_1^{-1}, & \mathbf{D}_1\mathbf{V}_1^*\mathbf{V}_2\\ 0 & 0\end{bmatrix}=\begin{bmatrix}\mathbf{D}_1 & 0\\ 0 & 0\end{bmatrix}$$

as required.

We now define the pseudo-inverse $\mathbf{A}^I$ of an $m\times n$ matrix $\mathbf{A}$ which has the singular decomposition (1) as the $n\times m$ matrix

$$\mathbf{A}^I=\mathbf{V}\begin{bmatrix}\mathbf{D}_1^{-1} & 0\\ 0 & 0\end{bmatrix}\mathbf{U}^*.$$

It is easy to verify that $\mathbf{A}^I$ satisfies the following axioms:

$$\mathbf{A}^I\mathbf{A}\mathbf{A}^I=\mathbf{A}^I \tag{4}$$

$$\mathbf{A}\mathbf{A}^I\mathbf{A}=\mathbf{A} \tag{5}$$

$$\mathbf{A}\mathbf{A}^I=(\mathbf{A}\mathbf{A}^I)^* \tag{6}$$

$$\mathbf{A}^I\mathbf{A}=(\mathbf{A}^I\mathbf{A})^* \tag{7}$$

and that when $\mathbf{A}$ is a non-singular square matrix then

$$\mathbf{A}^I=\mathbf{A}^{-1}.$$

Conversely it can be shown that the above four axioms of Moore—Penrose define the pseudo-inverse uniquely. Thus if $\mathbf{A}^I=\mathbf{A}_1$ and $\mathbf{A}^I=\mathbf{A}_2$ each satisfy

(4)—(7), we can prove that $\mathbf{A}_1 = \mathbf{A}_2$. In fact

$$
\begin{aligned}
\mathbf{A}\mathbf{A}_2 &= (\mathbf{A}\mathbf{A}_2)^* && \text{by (6) for } \mathbf{A}_2 \\
&= (\mathbf{A}\mathbf{A}_1\mathbf{A}\mathbf{A}_2)^* && \text{by (5) for } \mathbf{A}_1 \\
&= (\mathbf{A}\mathbf{A}_2)^*(\mathbf{A}\mathbf{A}_1)^* && \text{by properties of 'star'} \\
&= \mathbf{A}\mathbf{A}_2\mathbf{A}\mathbf{A}_1 && \text{by (6) for } \mathbf{A}_1 \text{ and } \mathbf{A}_2 \\
&= \mathbf{A}\mathbf{A}_1 && \text{by (5) for } \mathbf{A}_2.
\end{aligned} \tag{8}
$$

Similarly we establish

$$\mathbf{A}_1\mathbf{A} = \mathbf{A}_2\mathbf{A}. \tag{9}$$

We now have

$$
\begin{aligned}
\mathbf{A}_1 &= \mathbf{A}_1\mathbf{A}\mathbf{A}_1 && \text{by (4) for } \mathbf{A}_1 \\
&= (\mathbf{A}_2\mathbf{A})\mathbf{A}_1 && \text{by (9)} \\
&= \mathbf{A}_2(\mathbf{A}\mathbf{A}_1) && \\
&= \mathbf{A}_2\mathbf{A}\mathbf{A}_2 && \text{by (8)} \\
&= \mathbf{A}_2 && \text{by (4) for } \mathbf{A}_2.
\end{aligned}
$$

$$\boxed{A}\cdot\boxed{A^*}=\boxed{AA^*}; \quad \boxed{U^*}\cdot\boxed{AA^*}\cdot\boxed{U}=\underbrace{\begin{bmatrix} D_1^2 & 0 \\ 0 & 0 \end{bmatrix}}_{D};$$

$$\boxed{U^*}\cdot\boxed{A}=\boxed{F}; \quad \begin{bmatrix} F_1 \\ F_2=0 \end{bmatrix}\cdot\begin{bmatrix} F_1^* & F_2^* \end{bmatrix}=\begin{bmatrix} D_1^2 & 0 \\ 0 & 0 \end{bmatrix}$$

$$\underbrace{\begin{bmatrix} D_1^{-1}F_1=V_1^* \\ V_2^* \end{bmatrix}}_{V^*}\cdot\underbrace{\begin{bmatrix} V_1 & V_2 \end{bmatrix}}_{V}=\begin{bmatrix} V_1^*V_1 & V_1^*V_2 \\ V_2^*V_1 & V_2^*V_2 \end{bmatrix}=I_m$$

$$\boxed{U^*}\cdot\boxed{A}\cdot\boxed{V}=\underbrace{\begin{bmatrix} D_1 & 0 \\ 0 & 0 \end{bmatrix}}_{\Sigma}$$

$$\underbrace{\boxed{}}_{A^I}=\underbrace{\boxed{}}_{V^*}\cdot\underbrace{\begin{bmatrix} D_1^{-2} & 0 \\ 0 & 0 \end{bmatrix}}_{\Sigma^I}\cdot\underbrace{\boxed{}}_{U}$$

$$\boxed{V^*}\cdot\boxed{A^*}\cdot\boxed{A}\cdot\boxed{V}=\begin{bmatrix} D_1^2 & 0 \\ 0 & 0 \end{bmatrix}$$

Since the solution to $\mathbf{A}x=b$ in the case where $\mathbf{A}$ is non-singular is $x=\mathbf{A}^{-1}b$, it is natural to enquire about the significance of $x=\mathbf{A}^I b$ in the general case. We shall show that this gives the least squares solution of the system $\mathbf{A}x=b$. This matter is rather delicate (compare Problems 11.6, 11.7).

We begin by observing that transposing the relation (5) and using (6) gives

$$\mathbf{A}^* = (\mathbf{A}\mathbf{A}^I\mathbf{A})^* = \mathbf{A}^*(\mathbf{A}\mathbf{A}^I)^* = \mathbf{A}^*\mathbf{A}\mathbf{A}^I. \tag{10}$$

The following identity is readily established:

$$\begin{aligned}&\{\mathbf{A}\mathbf{P}+(\mathbf{I}-\mathbf{A}\mathbf{A}^I)\mathbf{Q}\}^*\{\mathbf{A}\mathbf{P}+(\mathbf{I}-\mathbf{A}\mathbf{A}^I)\mathbf{Q}\}=\\&=(\mathbf{A}\mathbf{P})^*(\mathbf{A}\mathbf{P})+\{(\mathbf{I}-\mathbf{A}\mathbf{A}^I)\mathbf{Q}\}^*(\mathbf{I}-\mathbf{A}\mathbf{A}^I)\mathbf{Q}\end{aligned} \tag{11}$$

where $\mathbf{P}$ is any $n\times p$ matrix and $\mathbf{Q}$ any $m\times p$ matrix. In fact all that is required is to verify that

$$(\mathbf{A}\mathbf{P})^*(\mathbf{I}-\mathbf{A}\mathbf{A}^I)\mathbf{Q}+\{(\mathbf{I}-\mathbf{A}\mathbf{A}^I)\mathbf{Q}\}^*(\mathbf{A}\mathbf{P})=0$$

and the first product cancels by use of (10) and the second by use of (6) and (5).

If we take $p=1$ so that $\mathbf{P}, \mathbf{Q}$ are column vectors x, $-b$ of dimensions n and m respectively we can use (11) to find

$$\begin{aligned}\|\mathbf{A}x-b\| &= \|\mathbf{A}(x-\mathbf{A}^I b)+(\mathbf{I}-\mathbf{A}\mathbf{A}^*)(-b)\|\\&=\|\mathbf{A}x-\mathbf{A}\mathbf{A}^I b\|+\|\mathbf{A}\mathbf{A}^I b-b\|\end{aligned} \tag{12}$$

where $\|\cdot\|$ indicates the euclidean vector norm. It follows from (12) that

$$\|\mathbf{A}x-b\| > \|\mathbf{A}(\mathbf{A}^I b)-b\| \tag{13}$$

unless

$$\mathbf{A}x=\mathbf{A}\mathbf{A}^I b. \tag{14}$$

If we replace $\mathbf{A}$ in (11) by $\mathbf{A}^I$ and make the appropriate specializations of $\mathbf{P}$ to $-b$ and $\mathbf{Q}$ to x and use the fact that $\mathbf{A}^{II}=\mathbf{A}$ we obtain

$$\|\mathbf{A}^I b+(\mathbf{I}-\mathbf{A}^I\mathbf{A})x\| = \|\mathbf{A}^I b\|+\|(\mathbf{I}-\mathbf{A}^I\mathbf{A})x\|. \tag{15}$$

It follows from (4) that if (14) holds then

$$\mathbf{A}^I\mathbf{A}x=\mathbf{A}^I(\mathbf{A}\mathbf{A}^I)b=\mathbf{A}^I b$$

so that (15) gives

$$\|x\| = \|\mathbf{A}^I b\|+\|x-\mathbf{A}^I b\|. \tag{16}$$

We have now established the following theorem.

Theorem 12.2. *In general $x=\mathbf{A}^I b$ is the unique best least squares solution of the system $\mathbf{A}x=b$; however, if there are other x which satisfy $\mathbf{A}x=\mathbf{A}\mathbf{A}^I b$ then all are best least squares solutions and $x=\mathbf{A}^I b$ is characterized among them by having the least norm.*

We conclude this chapter by pointing out that theory of the pseudo-inverse can be developed starting from the minimal properties just described. This approach has been carried out in detail by Peters and Wilkinson and is very appropriate in a numerical analysis context. We give an outline of their treatment; for details see Computer J. 13, 309—316 (1970).

If $\mathbf{A}$ is an $m \times n$ matrix of rank r then [cf. Problem 12.1] it can be factorized in the form

$$\mathbf{A} = \mathbf{B}\mathbf{C}$$

where $\mathbf{B}$ is an $m \times r$ matrix of rank r and $\mathbf{C}$ an $r \times n$ matrix of rank r. All factorizations of this form are given by

$$\mathbf{A} = (\mathbf{B}_0 \mathbf{Y}^{-1})(\mathbf{Y}\mathbf{C}_0)$$

where $\mathbf{A} = \mathbf{B}_0 \mathbf{C}_0$ is any factorization and $\mathbf{Y}$ is any non-singular $r \times r$ matrix.

It can be shown that

$$x = \mathbf{C}^*(\mathbf{C}\mathbf{C}^*)^{-1}(\mathbf{B}^*\mathbf{B})^{-1}\mathbf{B}^* b \tag{17}$$

is the minimal least squares solution of $\mathbf{A}x = b$ and

$$\mathbf{A}^I = \mathbf{C}^*(\mathbf{C}\mathbf{C}^*)^{-1}(\mathbf{B}^*\mathbf{B})^{-1}\mathbf{B}^*. \tag{18}$$

It is easy to verify that $\mathbf{A}^I$ does not depend on the choice of the factorization $\mathbf{A} = \mathbf{B}\mathbf{C}$ and that the axioms (4)—(7) are satisfied.

Chapter 12, Problems

12.1. Show that if $\mathbf{A}$ is an $m \times n$ matrix of rank r it can be represented in the form

$$\mathbf{A} = \mathbf{B}\mathbf{C}$$

where $\mathbf{B}$ is an $m \times r$ matrix and $\mathbf{C}$ and $r \times n$ matrix, both of rank r. Discuss the uniqueness of this representation.

12.2. Solve Problems 11.6, 11.7 using the methods of this chapter.

12.3. Find the singular value decomposition and the pseudo-inverse of

$$\mathbf{A} = \frac{1}{15}\begin{bmatrix} -13 & -16 \\ 2 & 14 \\ -22 & 4 \end{bmatrix} \quad \text{and of} \quad \mathbf{B} = \frac{1}{55}\begin{bmatrix} 54 & 28 \\ 74 & 18 \\ -21 & -72 \end{bmatrix}.$$

12.4. Find the singular value decomposition and the pseudo-inverse of

$$\mathbf{A} = \begin{bmatrix} 1 & 0 & 1 & 1 \\ 1 & 0 & -1 & 0 \\ 1 & 1 & 0 & 1 \end{bmatrix}.$$

12.5.(a) Show that if $\mathbf{A}=\mathbf{U}\Sigma\mathbf{V}^*$ is the singular value decomposition of an $n\times n$ matrix $\mathbf{A}$ then $\mathbf{U}\mathbf{V}^*$ is the unitary part of the polar decomposition of $\mathbf{A}$, i.e., $\mathbf{A}=(\mathbf{U}\mathbf{V}^*)\mathbf{H}$, where $\mathbf{H}$ is positive semidefinite hermitian.

(b) Show that if $\|\mathbf{A}\|_F=[\sum\sum|a_{ij}|^2]^{1/2}$ then

$$\|\mathbf{A}-\mathbf{U}\mathbf{V}^*\|_F\leqq\|\mathbf{A}-\mathbf{W}\|_F$$

for all unitary $\mathbf{W}$, i.e., $\mathbf{U}\mathbf{V}^*$ is the unitary matrix nearest to $\mathbf{A}$ in the sense of the Frobenius norm.

12.6. What are the pseudo-inverses of the following matrices:

(a) a column vector,

(b) a zero matrix?

Solutions to Selected Problems

1.7. Solution

Clearly

$$[2]^{-1}=[1/2], \quad \begin{bmatrix}3 & 1\\ 1 & 3\end{bmatrix}^{-1}=\frac{1}{8}\begin{bmatrix}3 & -1\\ -1 & 3\end{bmatrix}, \quad \begin{bmatrix}4 & 1 & 1\\ 1 & 4 & 1\\ 1 & 1 & 4\end{bmatrix}^{-1}=\frac{1}{18}\begin{bmatrix}5 & -1 & -1\\ -1 & 5 & -1\\ -1 & -1 & 5\end{bmatrix}$$

and the sum of the elements in the inverse matrix is $\frac{1}{2}$ in each case. We show that this is true in general.

The matrix $\mathbf{A}_3=n\mathbf{I}+\mathbf{J}$, where $\mathbf{J}$ is the matrix every element of which is 1, and in the special cases above the inverses are linear combinations of $\mathbf{I}$ and $\mathbf{J}$. Let us see if this is true in general. Assume $\mathbf{A}_3^{-1}=\alpha\mathbf{I}+\beta\mathbf{J}$. Then

$$(\alpha\mathbf{I}+\beta\mathbf{J})(n\mathbf{I}+\mathbf{J})=\mathbf{I}$$

which gives

$$n\alpha\mathbf{I}+(\alpha+\beta n)\mathbf{J}+\beta\mathbf{J}^2=\mathbf{I}$$

which can be satisfied, since $\mathbf{J}^2=n\mathbf{J}$, by taking

$$\alpha=\frac{1}{n}, \quad \beta=-\frac{\alpha}{2n}=-\frac{1}{2n^2}.$$

Since the inverse is unique we have

$$\mathbf{A}_3^{-1}=\frac{1}{n}\mathbf{I}-\frac{1}{2n^2}\mathbf{J}$$

and the sum of its elements is

$$n\times\frac{1}{n}+n^2\times\left(-\frac{1}{2n^2}\right)=1-\frac{1}{2}=\frac{1}{2}.$$

The answer in the case of the Hilbert matrix is n^2. See e.g. D. E. Knuth, *The art of computer programming*, I (1968), pp. 36/7, 473/4.

1.9. Solution

Since $R(\mathbf{x})=R(r\mathbf{x})$ for any $r\neq 0$ we may replace the condition $\mathbf{x}\neq 0$ by $\mathbf{x}'\mathbf{x}=1$. We know that we can choose an orthonormal system of vectors

$\boldsymbol{c}_1, \boldsymbol{c}_2, \ldots, \boldsymbol{c}_n$ which span the whole space $\boldsymbol{R}_n$ and which are characteristic vectors of $\boldsymbol{A}$, say $\boldsymbol{A}\boldsymbol{c}_i = \alpha_i \boldsymbol{c}_i$, $i=1, 2, \ldots, n$. Hence we can express any $\boldsymbol{x}$, with $\boldsymbol{x}'\boldsymbol{x}=1$ as

$$\boldsymbol{x} = \sum \xi_i \boldsymbol{c}_i,$$

where $\sum \xi_i^2 = 1$. Since

$$\begin{aligned}
\boldsymbol{x}'\boldsymbol{A}\boldsymbol{x} &= (\sum_i \xi_i \boldsymbol{c}_i)' \boldsymbol{A} (\sum_j \xi_j \boldsymbol{c}_j) \\
&= (\sum_i \xi_i \boldsymbol{c}_i')(\sum_j \xi_j \alpha_j \boldsymbol{c}_j) \\
&= \sum_{i,j} \alpha_j \xi_i \xi_j \boldsymbol{c}_i' \boldsymbol{c}_j \\
&= \sum_i \alpha_i \xi_i^2 \quad \text{(by orthonormality)}
\end{aligned}$$

we have

$$\alpha_n = \alpha_n \sum_i \xi_i^2 \leq R(\boldsymbol{x}) = \sum_i \alpha_i \xi_i^2 \leq \alpha_1 \sum_i \xi_i^2 = \alpha_1.$$

Also, clearly, for any i,

$$R(\boldsymbol{c}_i) = \alpha_i,$$

and so the bounds are attained.

In view of the importance of the Rayleigh quotient in numerical mathematics we add three remarks, the first two dealing with the two-dimensional case.

(1) We show how the Rayleigh quotient varies when $\boldsymbol{A} = \begin{bmatrix} a & h \\ h & b \end{bmatrix}$. By homogeneity we can restrict ourselves to vectors of unit length say $\boldsymbol{x} = \begin{bmatrix} \cos\theta \\ \sin\theta \end{bmatrix}$. Then

$$\begin{aligned}
Q(\theta) &= a\cos^2\theta + 2h\cos\theta\sin\theta + b\sin^2\theta \\
&= \tfrac{1}{2}[(a-b)\cos 2\theta + 2h\sin 2\theta + (a+b)].
\end{aligned}$$

To study the variation of $Q(\theta)$ with θ observe that

$$\begin{aligned}
q(\varphi) &= \alpha\cos\varphi + \beta\sin\varphi + \gamma \\
&= \sqrt{\alpha^2+\beta^2}\,\left[(\alpha/\sqrt{\alpha^2+\beta^2})\cos\varphi + (\beta/\sqrt{\alpha^2+\beta^2})\sin\varphi\right] + \gamma \\
&= \sqrt{\alpha^2+\beta^2}\,\sin(\varphi+\psi) + \gamma \quad \text{where } \sin\psi = \alpha/\sqrt{\alpha^2+\beta^2}, \ \cos\psi = \beta/\sqrt{\alpha^2+\beta^2},
\end{aligned}$$

and so $q(\varphi)$ oscillates between $\gamma \pm \sqrt{\alpha^2+\beta^2}$. Hence $Q(\theta)$ oscillates between (the two real numbers)

$$\tfrac{1}{2}(a+b) \pm \tfrac{1}{2}\sqrt{(a-b)^2 + 4h^2},$$

i.e., between the characteristic values of $\boldsymbol{A}$.

(2) The fact that the characteristic vectors are involved can be seen by use of the Lagrange Multipliers. To find extrema of $ax^2+2hxy+by^2$ subject to $x^2+y^2=1$, say, we compute E_x, E_y where

$$E=ax^2+2hxy+by^2-\lambda(x^2+y^2-1).$$

Then

$$E_x=2(a-\lambda)x+2hy, \quad E_y=2hx+2(b-\lambda)y$$

and at an extremum

$$\left.\begin{aligned}(a-\lambda)x+hy=0\\ hx+(b-\lambda)y=0\end{aligned}\right\}.$$

For a non-trivial solution we must have

$$\det\begin{bmatrix}a-\lambda & h\\ h & b-\lambda\end{bmatrix}=0,$$

i.e., λ must be a characteristic value of $\begin{bmatrix}a & h\\ h & b\end{bmatrix}$.

(3) A very important general principle should be pointed out here. At an extremum x_0 of $y=f(x)$ at which $f(x)$ is smooth, it is true that x "near" x_0 implies $f(x)$ "very near" $f(x_0)$. In the simplest case, $f(x)=x^2$ and $x_0=0$, we have $f(x)=x^2$ of the "second order" in x; this is not true if we do not insist on smoothness, as in shown by the case $g(x)=|x|$, $x_0=0$, in which $g(x)$ is of the same order as x. We are just using the Taylor expansion about x_0: $f(x)-f(x_0)=(x-x_0)^2\,[\frac{1}{2}f''(x_0)+\ldots]$ in the case where $f'(x_0)=0$.

This idea can be generalized to the case where $y=f(\mathbf{x})$ is a scalar function of a vector variable $\mathbf{x}$ in particular $y=R(\mathbf{x})$. It means that from a "good" guess at a characteristic vector of $\mathbf{A}$, the Rayleigh quotient gives a "very good" estimate of the corresponding characteristic value.

1.10. Solution

$$\begin{aligned}(\mathbf{I}-2\omega\omega')(\mathbf{I}-2\omega\omega')' &= (\mathbf{I}-2\omega\omega')(\mathbf{I}-2\omega\omega')\\ &= \mathbf{I}-4\omega\omega'+4\omega\omega'\omega\omega'\\ &= \mathbf{I}-4\omega\omega'+4\omega(\omega'\omega)\omega'\\ &= \mathbf{I}-4\omega\omega'+4\omega\omega'\\ &= \mathbf{I}.\end{aligned}$$

Matrices of the form of $\mathbf{0}$ were introduced by Householder and are of great use in numerical algebra. (See e.g. Chapter 8.)

Chapter 2

2.4. Solution

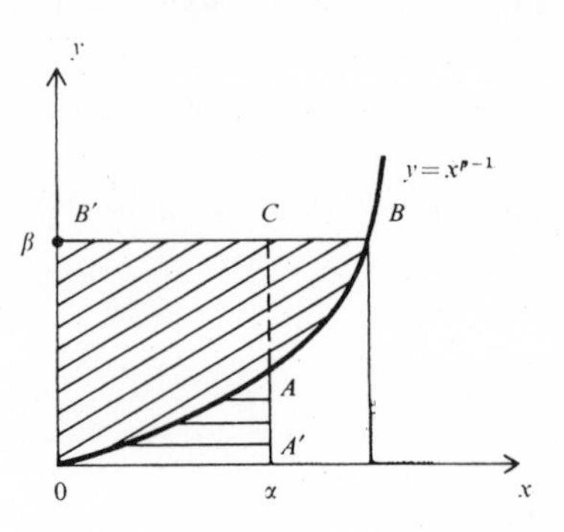

Assume $p > 1, \frac{1}{p} + \frac{1}{q} = 1, \alpha > 0, \beta > 0.$

Then area $0A'A = \int_0^\alpha x^{p-1}\,dx = \frac{\alpha^p}{p}$

and area $0BB' = \int_0^\beta y^{1/(p-1)}\,dy = \frac{\beta^q}{q}.$

Clearly the area of the rectangle $0A'CB'$ is not greater than the sum of the areas of the curvilinear triangles $0A'A$ and $0BB'$ and equal to it only if A, B and C coalesce. Hence

$$\alpha\beta \leqq \frac{\alpha^p}{p} + \frac{\beta^q}{q}$$

with strict inequality unless $\beta^q = \alpha^p$. This inequality, when written in the form

$$A^{1/p} B^{1/q} \leqq (A/p) + (B/q)$$

can be recognized as a generalization of the Arithmetic-Geometric Mean inequality

$$\sqrt{AB} \leqq (A+B)/2$$

from which it can be deduced, first when the weights p, q are rational and then by a limiting process for general p, q.

If we write $\alpha = \frac{|x_i|}{\|\boldsymbol{x}\|_p}$, $\beta = \frac{|y_i|}{\|\boldsymbol{y}\|_q}$ in this inequality we find

$$\text{(1)} \qquad \frac{|x_i|\,|y_i|}{\|\boldsymbol{x}\|_p \, \|\boldsymbol{y}\|_q} \leqq \frac{1}{p} \cdot \frac{|x_i|^p}{\sum |x_i|^p} + \frac{1}{q} \cdot \frac{|y_i|^q}{\sum |y_i|^q}.$$

Adding the last inequalities for $i = 1, 2, \ldots, n$ we find

$$\frac{\sum |x_i|\,|y_i|}{\|\boldsymbol{x}\|_p \, \|\boldsymbol{y}\|_q} \leqq \frac{1}{p} + \frac{1}{q} = 1,$$

so that

$$\text{(H)} \qquad \sum |x_i|\,|y_i| \leqq \|\boldsymbol{x}\|_p \, \|\boldsymbol{y}\|_q.$$

This is the Hölder inequality. There is equality in the last inequality if and only if there is equality in all the inequalities (1) which means that the $|x_i|^p$ are proportional to the $|y_i|^q$.

Observe that when $p=q=2$ the inequality (H) reduces to the Schwarz inequality

(S) $$[\sum |x_i|\,|y_i|]^2 \leqq \sum |x_i|^2 \sum |y_i|^2.$$

Observe also that the limiting case of (H), when $p=1$, $q=\infty$, is also valid.

In order to establish the Minkowski inequality

(M) $$\|\boldsymbol{x}+\boldsymbol{y}\|_p \leqq \|\boldsymbol{x}\|_p + \|\boldsymbol{y}\|_p$$

we write

$$(|x_i|+|y_i|)^p = |x_i|\,(|x_i|+|y_i|)^{p-1} + |y_i|\,(|x_i|+|y_i|)^{p-1}$$

and sum, applying (H) twice on the right to get

$$\sum (|x_i|+|y_i|)^p \leqq \|\boldsymbol{x}\|_p \left[\sum (|x_i|+|y_i|)^{(p-1)q}\right]^{1/q} + \|\boldsymbol{y}\|_p \left[\sum (|x_i|+|y_i|)^{(p-1)q}\right]^{1/q}.$$

Observe that $(p-1)q=p$, so that the terms in [ ] on the right are identical with that on the left. Hence, dividing through,

$$\left[\sum (|x_i|+|y_i|)^p\right]^{1-(1/q)} \leqq \|\boldsymbol{x}\|_p + \|\boldsymbol{y}\|_p,$$

i.e., since $1-(1/q)=1/p$,

$$\|\boldsymbol{x}+\boldsymbol{y}\|_p \leqq \|\boldsymbol{x}\|_p + \|\boldsymbol{y}\|_p.$$

The equality cases can easily be distinguished.

We have therefore shown that the p-norm satisfies Axiom 3, the triangle-inequality. The proofs that Axioms 1, 2 are satisfied are trivial.

To complete the solution we observe that

$$(\max |x_i|)^p \leqq \|\boldsymbol{x}\|_p^p = \sum |x_i|^p \leqq n(\max |x_i|)^p$$

which we can write as

$$\|\boldsymbol{x}\|_\infty^p \leqq \|\boldsymbol{x}\|_p^p \leqq n\|\boldsymbol{x}\|_\infty^p.$$

Taking p-th roots we get

$$\|\boldsymbol{x}\|_\infty \leqq \|\boldsymbol{x}\|_p \leqq n^{1/p}\,\|\boldsymbol{x}\|_\infty$$

and, since as $p\to\infty$,

$$n^{1/p}\to 1$$

we have

$$\|\boldsymbol{x}\|_\infty \leqq \lim_{p\to\infty} \|\boldsymbol{x}\|_p \leqq \|\boldsymbol{x}\|_\infty.$$

2.5. Solution

See sketch. For simplicity we have only drawn the part in the first quadrant. Each set is bounded, closed, convex and symmetrical about the origin ("equilibrated") and has a not-empty interior.

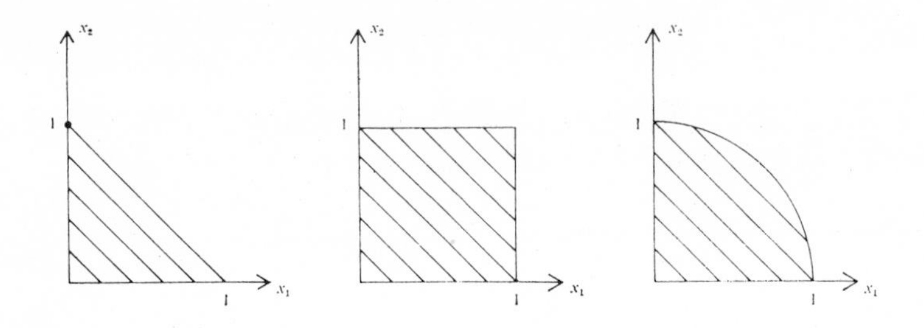

2.6. Solution

See sketch. For simplicity we have only drawn the part in the first quadrant. This set $\|\boldsymbol{x}\| \leqq 1$ is not convex but has the other properties of those in Problem 2.5. The triangle inequality is not satisfied: e.g.,

$$\boldsymbol{x} = [0, 1]', \quad \boldsymbol{y} = [1, 0]', \quad \boldsymbol{x} + \boldsymbol{y} = [1, 1]'$$

$$\|\boldsymbol{x} + \boldsymbol{y}\| = 2^{3/2} > 2 = \|\boldsymbol{x}\| + \|\boldsymbol{y}\|.$$

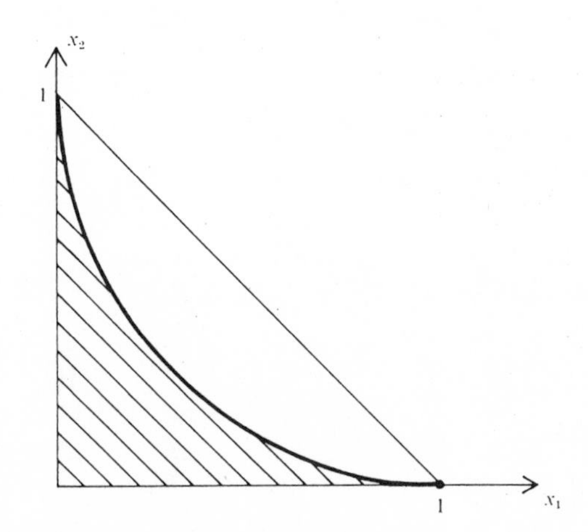

2.7. Solution

See sketch. For simplicity we have only drawn the part in the first quadrant.

The set $\|\boldsymbol{x}\| \leqq 1$ has the properties listed in Problem 2.5 and the axioms are satisfied.

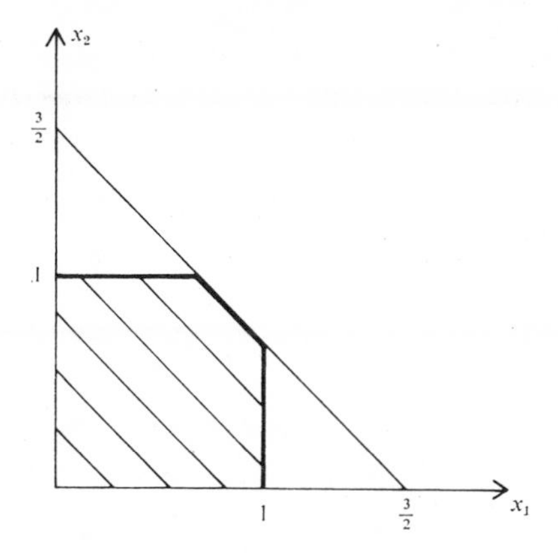

2.8. Solution

If p_1 and p_2 are equivalent and if p_2 and p_3 are equivalent then p_1 and p_3 are equivalent for we have

$$0 < p_{12}\, p_{23} \leq \frac{p_1(\boldsymbol{x})}{p_2(\boldsymbol{x})} \cdot \frac{p_2(\boldsymbol{x})}{p_3(\boldsymbol{x})} = \frac{p_1(\boldsymbol{x})}{p_3(\boldsymbol{x})} \leq p_{32}\, p_{21} < \infty.$$

It will be enough, therefore, to prove that any norm $p(\boldsymbol{x})$ is equivalent, e.g., to the Chebyshev norm $Ч(\boldsymbol{x}) = \|\boldsymbol{x}\|_\infty$. The set $S = \{\boldsymbol{x}\colon Ч(\boldsymbol{x}) = 1\}$, the surface of the appropriate cube, is closed and bounded. Any norm $p(\boldsymbol{x})$ is continuous everywhere. Let m, M be its lower and upper bounds on S, so that

$$m \leq p(\boldsymbol{x}) \leq M, \quad \boldsymbol{x} \in S.$$

Now, by continuity there are vectors $\boldsymbol{x}_m$, $\boldsymbol{x}_M$ in S such that $p(\boldsymbol{x}_m) = m$, $p(\boldsymbol{x}_M) = M$ and, since $\|\boldsymbol{x}_m\| = 1$, $m > 0$ and we have

$$0 < m \leq M < \infty.$$

For any vector $\boldsymbol{x} \neq 0$ there is a k such that $\boldsymbol{x} = k\xi$, where $Ч(\xi) = 1$. We have, therefore,

$$\frac{p(\boldsymbol{x})}{Ч(\boldsymbol{x})} = \frac{p(k\xi)}{Ч(k\xi)} = \frac{|k|\, p(\xi)}{|k|\, Ч(\xi)} = p(\xi).$$

Hence, for $\boldsymbol{x} \neq 0$,

$$0 < m \leq \frac{p(\boldsymbol{x})}{Ч(\boldsymbol{x})} \leq M < \infty.$$

It is instructive to deal with the two-dimensional case of the second part geometrically, drawing the contour lines of the norm surfaces $z = p(x_1, x_2)$. By homogeneity, the ratio $\|\mathbf{X}\|_2 / \|\mathbf{X}\|_1$ is equal to its value at the vectors $\mathbf{X}_1$, $\mathbf{X}_2$ where these are chosen to make $\|\mathbf{X}_1\|_1 = 1$, $\|\mathbf{X}_2\|_1 = \sqrt{2}$. Since $\mathbf{X}_1$ is inside the circle $\|x\|_2=1$ we have $\|\mathbf{X}_1\|_2 \leqq 1$ so that $\|\mathbf{X}_1\|_2 \leqq 1 = \|\mathbf{X}_1\|_1$ and then $\|\mathbf{X}\|_2 \leqq \|\mathbf{X}\|_1$. Since $\mathbf{X}_2$ is outside the circle $\|x\|_2=1$ we have $\|\mathbf{X}_2\|_2/\|\mathbf{X}_2\|_1 \geqq 1/\sqrt{2}$ and so $\sqrt{2}\|\mathbf{X}\|_2 \geqq \|\mathbf{X}_1\|_1$.

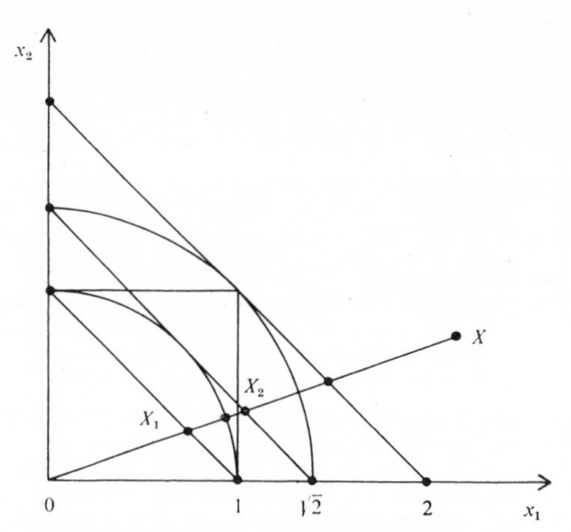

We deal with the general case analytically. We have $n \max |x_i| \geqq \sum |x_i| \geqq \geqq \max |x_i|$ which gives $n\|x\|_\infty \geqq \|x\|_1 \geqq \|x\|_\infty$.

Again $\|x\|_2 = \sqrt{(\sum |x_i|^2)} \leqq \sqrt{((\sum |x_i|)^2)} = \|x\|_1$ and, by the Schwarz inequality,

$$\|x\|_1 = \sum 1 \cdot |x_i| \leqq \sqrt{\{(\sum 1^2)(\sum |x_i|^2)\}} = \sqrt{n}\,\|x\|_2.$$

Finally $\|x\|_2 = \sqrt{(\sum |x_i|^2)} \geqq \sqrt{((\max |x_i|)^2)} = \|x\|_\infty$ and

$$\|x\|_2 = \sqrt{(\sum |x_i|^2)} \leqq \sqrt{(n(\max |x_i|)^2)} = \sqrt{n}\,\|x\|_\infty.$$

The equality cases in each inequality can be easily distinguished. All the result can be obtained from the fact that if $\infty \geqq q \geqq p \geqq 1$ we have

$$\|x\|_q \leqq \|x\|_p \leqq n^{p^{-1}-q^{-1}}\|x\|_q$$

and there is strict inequality on the right except when x is a unit vector and on the left except when $x=e=[1, 1, \ldots, 1]'$. This result can be obtained from the Hölder inequality. (See, e.g., G. H. Hardy, J. E. Littlewood and G. Pólya, *Inequalities*, or N. Gastinel, *Analyse numérique linéaire*, p. 29.)

2.9. Solution

We shall only discuss the Frobenius norm. It is trivial to check that the first two axioms are satisfied and we deal only with the last two.

Subadditivity:

In this paragraph all sums are double sums with respect to i, j.

$$\|\mathbf{A}+\mathbf{B}\|^2 = \sum (|a_{ij}+b_{ij}|)^2 \leqq \sum (|a_{ij}|^2 + |b_{ij}|^2 + 2|a_{ij}|\,|b_{ij}|)$$
$$\leqq \sum |a_{ij}|^2 + \sum |b_{ij}|^2 + 2\sum |a_{ij}|\,|b_{ij}|.$$

The Schwarz inequality gives

$$2\sum |a_{ij}|\,|b_{ij}| \leqq 2\sqrt{\sum |a_{ij}|^2}\sqrt{\sum |b_{ij}|^2}$$

which can be used in our last inequality to get

$$\|\mathbf{A}+\mathbf{B}\|^2 \leqq \|\mathbf{A}\|^2+\|\mathbf{B}\|^2+2\|\mathbf{A}\|\,\|\mathbf{B}\| = (\|\mathbf{A}\|+\|\mathbf{B}\|)^2$$

so that

$$\|\mathbf{A}+\mathbf{B}\| \leqq \|\mathbf{A}\|+\|\mathbf{B}\|$$

as required.
Submultiplicativity:

$$\begin{aligned}
\|\mathbf{A}\mathbf{B}\|^2 &= \sum_i \sum_j \left(\left|\sum_k a_{ik} b_{kj}\right|\right)^2 \\
&\leqq \sum_i \sum_j \left(\sum_k |a_{ik}|\,|b_{kj}|\right)^2 \\
&\leqq \sum_i \sum_j \sum_k |a_{ik}|^2 \sum_l |b_{lj}|^2 \quad \text{(by the Schwarz inequality)} \\
&= \left(\sum_i \sum_k |a_{ik}|^2\right)\left(\sum_j \sum_l |b_{lj}|^2\right) = \|\mathbf{A}\|^2\,\|\mathbf{B}\|^2
\end{aligned}$$

as required.

2.10. Solution

We note that $\operatorname{tr}\mathbf{A} = \sum a_{ii} = \sum \alpha_i$ and so

$$\operatorname{tr}\mathbf{A} = \operatorname{tr}\mathbf{S}^{-1}\mathbf{A}\mathbf{S}$$

for any non-singular $\mathbf{S}$, since $\mathbf{S}^{-1}\mathbf{A}\mathbf{S}$ and $\mathbf{A}$ have the same characteristic values.

It is clear that

$$\|\mathbf{A}\|_F^2 = \sum\sum a_{ij}^2 = \operatorname{tr}\mathbf{A}\mathbf{A}'.$$

The required result is established as follows:

$$\begin{aligned}
\|\mathbf{O}'\mathbf{A}\mathbf{O}\|_F^2 &= \operatorname{tr}(\mathbf{O}'\mathbf{A}\mathbf{O}(\mathbf{O}'\mathbf{A}\mathbf{O})') \\
&= \operatorname{tr}(\mathbf{O}'\mathbf{A}\mathbf{O}\mathbf{O}'\mathbf{A}'\mathbf{O}) \\
&= \operatorname{tr}\mathbf{O}'\mathbf{A}\mathbf{A}'\mathbf{O} \\
&= \operatorname{tr}\mathbf{A}\mathbf{A}' \quad \text{since } \mathbf{A}\mathbf{A}' \text{ and } \mathbf{O}'\mathbf{A}\mathbf{A}'\mathbf{O} \text{ are similar} \\
&= \|\mathbf{A}\|_F^2.
\end{aligned}$$

More indeed is true: if $\mathbf{O}_1$, $\mathbf{O}_2$ are orthogonal

$$\|\mathbf{O}_1\mathbf{A}\mathbf{O}_2\|_F^2 = \operatorname{tr}(\mathbf{O}_1\mathbf{A}\mathbf{O}_2\mathbf{O}_2'\mathbf{A}'\mathbf{O}_1') = \operatorname{tr}(\mathbf{O}_1\mathbf{A}\mathbf{A}'\mathbf{O}_1') = \|\mathbf{A}\|_F^2.$$

2.12. Solution

This terminology appears to be due to F. L. Bauer (in his lectures at Stanford University in 1967). Manhattan is the central borough of the city of New York and its streets and avenues form a rectangular grid. So the "taxi cab" distance between two points is the sum of the distances along the streets and along the avenues.

Chapter 3

3.2. Solution

The characteristic polynomial of $\mathbf{A}$ is x^2-5x-2 and the characteristic values are $\frac{1}{2}(5\pm\sqrt{33})$. Hence $\varrho(\mathbf{A})=\frac{1}{2}(5+\sqrt{33})$.

Chebyshev case. Clearly $\|\mathbf{A}\|=\max(3, 7)=7$. We assume $\|\boldsymbol{x}\|_\infty=1$ so that $\boldsymbol{x}'=[x_1, x_2]$ where $\max(|x_1|, |x_2|)=1$. Then

$$\|\mathbf{A}\boldsymbol{x}\|_\infty=\max(|x_1+2x_2|, |3x_1+4x_2|)\leqq 3|x_1|+4|x_2|\leqq 7$$

and we can have equality if and only if $|x_1|=|x_2|=1$.

Manhattan case. Clearly $\|\mathbf{A}\|=\max(4,6)=6$. We assume $|x_1|+|x_2|=1$. Then

$$\|\mathbf{A}\boldsymbol{x}\|_1=|x_1+2x_2|+|3x_1+4x_2|\leqq 4|x_1|+6|x_2|\leqq 6(|x_1|+|x_2|)=6,$$

and we can have equality if and only if $x_1=0$, $|x_2|=1$.

Euclidean case. We have $\mathbf{A}\mathbf{A}'=\begin{bmatrix}5 & 11\\ 11 & 25\end{bmatrix}$, and this matrix has for its dominant characteristic value $15+\sqrt{221}$ so that $\|\mathbf{A}\|=\sqrt{15+\sqrt{221}}$. We assume $x_1^2+x_2^2=1$. Then

$$\|\mathbf{A}\boldsymbol{x}\|_2^2=(x_1+2x_2)^2+(3x_1+4x_2)^2$$

$$=10x_1^2+28x_1x_2+20x_2^2.$$

In order that this should be $\|\mathbf{A}\|^2$, $\boldsymbol{x}$ must be the dominant characteristic vector of $\mathbf{A}\mathbf{A}'$ so that

$$10x_1+14x_2=(15+\sqrt{221})\,x_1$$

$$14x_1+20x_2=(15+\sqrt{221})\,x_2.$$

These equations are necessarily consistent and we find

$$|x_1| : |x_2| = 14 : 5+\sqrt{221}.$$

3.3. Solution

The fact that the matrix norm induced by a vector norm is multiplicative depends essentially on the remark that

$$\frac{n(\mathbf{A}\mathbf{B}\boldsymbol{x})}{n(\boldsymbol{x})}=\frac{n(\mathbf{A}(\mathbf{B}\boldsymbol{x}))}{n(\mathbf{B}\boldsymbol{x})}\cdot\frac{n(\mathbf{B}\boldsymbol{x})}{n(\boldsymbol{x})}$$

and this does not apply in the mixed case.

To see that a mixed norm is not necessarily submultiplicative take, as we may, $n_2(\boldsymbol{x})=kn_1(\boldsymbol{x})$ where k is arbitrary. Then

$$\|\mathbf{A}\|_{12}=\sup\frac{n_1(\mathbf{A}\boldsymbol{x})}{kn_1(\boldsymbol{x})}=\frac{1}{k}\|\mathbf{A}\|_{11}$$

where $\|\mathbf{A}\|_{11}$ is the matrix norm induced by the vector norm $n_1(\boldsymbol{x})$.

We know that

$$\|\mathbf{AB}\|_{11}\leqq\|\mathbf{A}\|_{11}\|\mathbf{B}\|_{11}$$

so that

$$\|\mathbf{AB}\|_{12}\leqq k^{-1}\|\mathbf{A}\|_{12}\|\mathbf{B}\|_{12}$$

and we can certainly choose k so that this is false.

It is easy to show that if $m(\mathbf{A})$ is not submultiplicative, then we can choose a multiplier μ so that $\mu m(\mathbf{A})$ is a submultiplicative norm. In fact, the arguments of Problem 2.8 show also that any two matrix norms (finite-dimensional) are equivalent. Hence there are positive constants a, b such that

$$a\|\mathbf{A}\|_\infty\leqq m(\mathbf{A})\leqq b\|\mathbf{A}\|_\infty.$$

Hence

$$m(\mathbf{AB})\leqq b\|\mathbf{AB}\|_\infty\leqq b\|\mathbf{A}\|_\infty\|\mathbf{B}\|_\infty\leqq\frac{b}{a^2}m(\mathbf{A})m(\mathbf{B}).$$

Thus if we take $\mu=b/a^2$ then

$$\mu m(\mathbf{AB})\leqq\mu m(\mathbf{A})\times\mu m(\mathbf{B}).$$

Corresponding to any not submultiplicative matrix norm $m(\mathbf{A})$ there is a least multiplier $\mu\geqq1$ such that $\mu M(\mathbf{A})$ is submultiplicative.

In the case when $m(\mathbf{A})=\max\limits_{i,j}|a_{i,j}|$ it can be shown that the least multiplier is n.

(See J. F. Maitre, Numer. Math., *10*, 132—161 (1967).)

The last part of the problem, when put in standard notation, requires us to find

$$m_{\infty,1}=\sup_{x\neq0}\frac{\|\mathbf{A}\boldsymbol{x}\|_\infty}{\|\boldsymbol{x}\|_1}.$$

Clearly

$$m_{\infty,1}=\sup_{x\neq0}\frac{\max\limits_i\sum\limits_{j=1}^n|a_{ij}x_j|}{\sum|x_j|}\leqq\sup_{x\neq0}\frac{\max\limits_{i,j}|a_{ij}|\sum\limits_{j=1}^n|x_j|}{\sum|x_j|}=\max_{i,j}|a_{ij}|.$$

Hence $m_{\infty,1}\leqq\max\limits_{i,j}|a_{ij}|$. We prove $m_{\infty,1}\geqq\max\limits_{i,j}|a_{ij}|$ and so $m_{\infty,1}=\max\limits_{i,j}|a_{ij}|$. In fact if $|a_{IJ}|=\max|a_{ij}|$ then if

$\boldsymbol{x}=\boldsymbol{e}_J$ we have $\mathbf{A}\boldsymbol{x}=[a_{1J},a_{2J},\dots,a_{nJ}]'$ and $\|\mathbf{A}\boldsymbol{x}\|_\infty=|a_{IJ}|$, $\|\boldsymbol{x}\|_1=1$.

The values of $m_{\infty,2}$, $m_{1,2}$ do not seem to be known.

3.5. Solution

To prove $\|\mathbf{A}\,\boldsymbol{x}\|_2 \leqq \|\mathbf{A}\|_F\,\|\boldsymbol{x}\|_2$ observe that

$$\|\mathbf{A}\,\boldsymbol{x}\|_2^2 = \sum_i \left|\sum_j a_{ij}\,x_j\right|^2 \leqq \sum_i \sum_j |a_{ij}|^2 \sum_k |x_k|^2 \text{ (by Schwarz)}$$

$$= \|\mathbf{A}\|_F^2\,\|\boldsymbol{x}\|_2^2$$

and take square roots.

It is clear that

$$\|\mathbf{A}\|_F^2 = \sum_i \sum_j |a_{ij}|^2 \leqq n^2 \max_{i,\,j} |a_{ij}|^2 = \|\mathbf{A}\|_M^2$$

and so, a fortiori,

$$\|\mathbf{A}\,\boldsymbol{x}\|_2 \leqq \|\mathbf{A}\|_M\,\|\boldsymbol{x}\|_2 .$$

3.6. Solution

The fact that $n(\boldsymbol{x})$ is a norm is easy. Clearly $m \geqq 0$ implies $n \geqq 0$ and $\boldsymbol{x}\boldsymbol{\eta}'=0$ if and only if $\boldsymbol{x}=0$ and so $n(\boldsymbol{x})=m(\boldsymbol{x}\boldsymbol{\eta}')=0$ if and only if $\boldsymbol{x}=0$. Homogeneity of m implies homogeneity of n. For the triangle inequality

$$n(\boldsymbol{x}+\boldsymbol{y}) = m\big((\boldsymbol{x}+\boldsymbol{y})\boldsymbol{\eta}'\big) = m(\boldsymbol{x}\boldsymbol{\eta}'+\boldsymbol{y}\boldsymbol{\eta}')$$

$$\leqq m(\boldsymbol{x}\boldsymbol{\eta}') + m(\boldsymbol{y}\boldsymbol{\eta}')$$

$$= n(\boldsymbol{x}) + n(\boldsymbol{y}).$$

To prove compatibility we want to show

$$n(\mathbf{A}\,\boldsymbol{x}) \leqq m(\mathbf{A})\,n(\boldsymbol{x}),$$

i.e.,

$$m(\mathbf{A}\,\boldsymbol{x}\boldsymbol{\eta}') \leqq m(\mathbf{A})\,m(\boldsymbol{x}\boldsymbol{\eta}'),$$

but this is just the assertion of submultiplicativity of m.

3.7. Solution

Let $N=\{\boldsymbol{x}|n(\boldsymbol{x})=1\}$; this is regarded as a subset of the euclidean plane $\mathbf{R}_2$ and closed and bounded are relative to the geometry of $\mathbf{R}_2$.

A set $S \subset \mathbf{R}_2$ is *bounded* if there is a constant M such that if $[x, y]' \in S$, $x^2+y^2 \leqq M$.

A set $S \subset \mathbf{R}_2$ is *closed* if whenever a sequence of points $s_n \in S$ has a limit s, i.e., if $\lim\,[(x_n-x)^2+(y_n-y)^2]=0$, then $s \in S$.

The set of points U on the circumference of the unit circle $x^2+y^2=1$ is manifestly bounded (take $M=1$) and can be proved to be closed as follows. Suppose $\lim s_n = s$ — this means $\lim\,[(x_n-x)^2+(y_n-y)^2]=0$ which implies $\lim x_n = x$, $\lim y_n = y$. Since $x_n^2+y_n^2=1$ we have $\lim x_n^2 + \lim y_n^2 = 1$, both limits existing; but this is just $x^2+y^2=1$.

We now note that any norm is continuous. From the triangle inequality,

$$\begin{aligned} \big|\|z+\zeta\| - \|z\|\big| &\leq \|\zeta\| \leq |\xi|\,\|e_1\| + |\eta|\,\|e_2\| \quad \text{if} \quad \zeta = \xi e_1 + \eta e_2 \\ &\leq c\{|\xi| + |\eta|\} \quad \text{if} \quad c = \max(\|e_1\|, \|e_2\|) \\ &\leq c\sqrt{2}\cdot\sqrt{|\xi|^2+|\eta|^2}\,, \text{ by Schwarz.} \end{aligned}$$

Thus if $|\xi|^2+|\eta|^2$ is small so is $\big|\|z+\xi\|-\|z\|\big|$.

(1) *N is bounded.* If not there is a sequence of points (or vectors) $z_n \in N$ such that $x_n^2+y_n^2 \to \infty$. Consider the sequence of points $\zeta^{(n)} = z_n/\sqrt{x_n^2+y_n^2}$. These lie on U and by the Bolzano—Weierstrass theorem have a limit point $\zeta \in U$ (since U is closed). Suppose $\zeta^{(n_i)} \to \zeta$. Then $\|\zeta^{(n_i)}\| = \dfrac{1}{\sqrt{x_{n_i}^2+y_{n_i}^2}}$ by homogeneity; but this tends to zero. By continuity of norm $\|\zeta\|=0$ and $\zeta \neq 0$ — a contradiction with the first norm axiom.

(2) *N is closed.* If $z_n \in N$ and $z_n \to z$ then by continuity of the norm $1 = \lim \|z_n\| = \|\lim z_n\| = \|z\|$ so that $z \in N$.

3.8. Solution

The absolute row sums are 3, 4, 4, 3. Hence $\|\mathbf{A}\|_\infty = 4$. The maximizing row is $[1, -2, 1, 0]$ and so an extremal vector is

$$\boldsymbol{x}_0 = \begin{bmatrix} 1 \\ -1 \\ 1 \\ 0 \end{bmatrix} \quad \text{which gives} \quad \mathbf{A}\boldsymbol{x}_0 = \begin{bmatrix} -3 \\ 4 \\ -3 \\ 1 \end{bmatrix} \quad \text{with} \quad \|\boldsymbol{x}_0\|_\infty = 1, \quad \|\mathbf{A}\boldsymbol{x}_0\|_\infty = 4.$$

The matrix $\mathbf{A}_8$ is symmetric. Hence $\mathbf{A}'\mathbf{A} = \mathbf{A}^2$. If $\boldsymbol{a}, \alpha$ is a characteristic pair for $\mathbf{A}$ so that $\mathbf{A}\boldsymbol{a} = \alpha\boldsymbol{a}$ we have $\mathbf{A}^2\boldsymbol{a} = \mathbf{A}(\mathbf{A}\boldsymbol{a}) = \mathbf{A}(\alpha\boldsymbol{a}) = \alpha(\mathbf{A}\boldsymbol{a}) = \alpha^2\boldsymbol{a}$, i.e., the characteristic values of $\mathbf{A}^2$ are the squares of those of $\mathbf{A}$. Hence $\|\mathbf{A}\|_2 = \varrho(\mathbf{A})$.

Now from Problem 5.13 (iv), or by direct verification, the characteristic values of $\mathbf{A}$ are $\alpha_r = -2+2\cos(r\pi/5)$, $r = 1, 2, 3, 4$. The spectral radius of $\mathbf{A}$ corresponds to $r=4$ and $\varrho(\mathbf{A}) = -\alpha_4 = 2+2\cos(\pi/5) = \frac{1}{2}(5+\sqrt{5})$. The dominant characteristic vector of $\mathbf{A}$ is $\boldsymbol{a}_4 = [1, -2c, 2c, 1]'$ where $c = \cos(\pi/5)$ and this gives equality since $\mathbf{A}\boldsymbol{a} = \alpha\boldsymbol{a}$ so that $\|\mathbf{A}\boldsymbol{a}_4\| = |\alpha|\,\|\boldsymbol{a}_4\| = \varrho(\mathbf{A})\,\|\boldsymbol{a}_4\|$.

The absolute column sums of $\mathbf{A} = \mathbf{A}_{16}$ are 4, 9, 9, 4. Hence $\|\mathbf{A}\|_1 = 9$. The extremal vector is $\boldsymbol{x}_0 = [0, 1, 0, 0]'$ for which $\mathbf{A}\boldsymbol{x}_0 = [0, 2, 3, 4]'$ and $\|\boldsymbol{x}_0\|_1 = 1$, $\|\mathbf{A}\boldsymbol{x}_0\| = 2+3+4 = 9$.

Chapter 4

4.1. Solution

We approach the problem in the following way. To invert $\mathbf{A}$ we have to solve the systems of equations

$$\mathbf{A}c_i = e_i$$

where c_i is the i-th column of $\mathbf{A}^{-1}$ and e_i the i-th unit vector.

Assume, for simplicity, that the triangularization can be carried out without rearrangements. This requires, in the first stage, the determination of the factors $-\dfrac{a_{21}}{a_{11}}, -\dfrac{a_{31}}{a_{11}}, \ldots, -\dfrac{a_{n1}}{a_{11}}$ and then the multiplication of the first row by $-a_{r1}/a_{11}$ and its addition to the r-th row, so as to kill the $(r, 1)$ term.

We need about n multiplications and then a further $(n-1)\times(n-1)$, in all about n^2.

For the whole process we need about $\sum r^2 = n^3/3$.

Let us now consider these operations carried out on the right hand sides. Take the case of e_i. No action is needed until the i-th stage when we add multiples of 1 to the zeros in the $i+1, \ldots, n$-th position. At the next stage we have to add multiples of the $(i+1)$-st component to those in the $i+2, \ldots, n$-th position. In all we need about

$$\sum_{r=i}^{n} (n-r) \doteqdot \frac{(n-i)^2}{2}$$

multiplications. To deal with all the right hand sides therefore requires about

$$\sum_{i=1}^{n} \frac{(n-i)^2}{2} \doteqdot \sum_{j=1}^{n} \frac{j^2}{2} \doteqdot \frac{n^3}{6}$$

multiplications.

Finally we have to solve n triangular systems: each involves about $n^2/2$ multiplications and so, in all about $n^3/2$ multiplications.

The grand total is $\dfrac{n^3}{3} + \dfrac{n^3}{6} + \dfrac{n^3}{2} = n^3$.

4.2. Solution

The factorization of an $(m+1)\times(m+1)$ matrix which we can regard as a band matrix of width $2m+1$ gives (in general) full triangular factors, which we can regard as band matrices.

Suppose we have established, for $n \geqq m+1$ that an $n\times n$ band matrix of width $2m+1$ can be factorized into matrices of the same type. Consider now the problem for an $(n+1)\times(n+1)$ matrix, using the notation of p. 33 and the assumption that all the leading submatrices of $\mathbf{A}$ are non-singular.

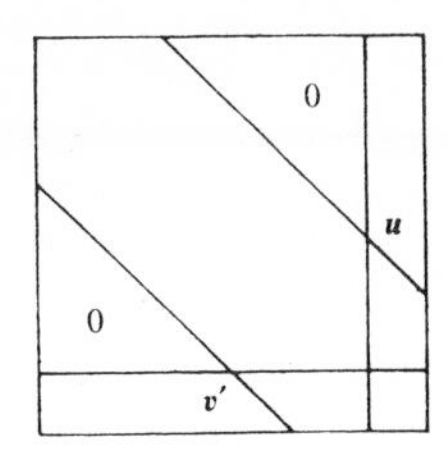

=

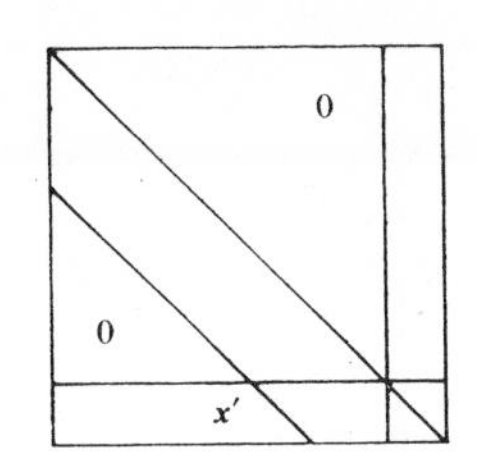

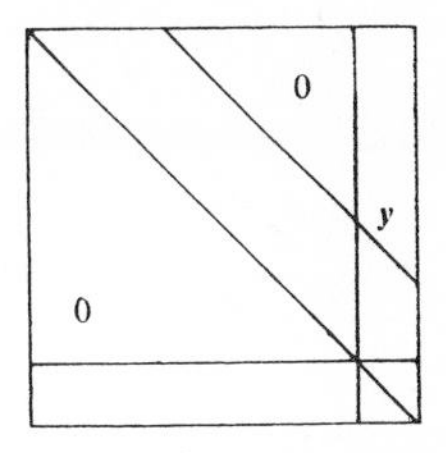

Our induction hypothesis validates the diagram above. We have only to show that the vectors $\boldsymbol{x}'$, $\boldsymbol{y}$ will each have $n-m$ initial zeros. These are obtained as

$$\boldsymbol{x}' = \boldsymbol{v}' \mathbf{U}_{n-1}^{-1}, \quad \boldsymbol{y} = \mathbf{L}_{n-1}^{-1} \boldsymbol{u}$$

and since $\boldsymbol{v}$, $\boldsymbol{u}$ have $n-m$ initial zeros so have $\boldsymbol{x}$, $\boldsymbol{y}$.

4.3. Solution

If $a_{11} \neq 0$ we obtain the first Gauss transform of $\mathbf{A}$ by killing the remaining elements in the first row by premultiplication by

$$\mathbf{L}_1 = \begin{bmatrix} 1 & & & \\ -a_{21}/a_{11} & 1 & & \\ \vdots & & \ddots & \\ -a_{n1}/a_{11} & \dots & & 1 \end{bmatrix}.$$

We write $\mathbf{L}_1 \mathbf{A} = [a_{ij}^{(2)}]$ and for convenience $\mathbf{A} = [a_{ij}^{(1)}]$. We denote by $[\mathbf{A}]_r$ the leading $r \times r$ submatrix of $\mathbf{A}$:

$$[\mathbf{A}]_r = \begin{bmatrix} a_{11} & \dots & a_{1r} \\ \vdots & & \vdots \\ a_{r1} & \dots & a_{rr} \end{bmatrix}$$

so that

$$\det [\mathbf{A}]_1 = a_{11}, \quad \det [\mathbf{A}]_2 = a_{11} a_{22} - a_{12} a_{21}, \dots .$$

It is clear that

$$a_{22}^{(2)} = a_{22} - \frac{a_{12} a_{21}}{a_{11}} = \frac{\det [\mathbf{A}]_2}{\det [\mathbf{A}]_1}.$$

If $a_{22}^{(2)}$ is not zero we can get the second Gauss transform of $\mathbf{A}$ by premultiplying $\mathbf{L}_1 \mathbf{A}$ by

$$\mathbf{L}_2 = \begin{bmatrix} 1 & & & & \\ & 1 & & & \\ & -a_{32}^{(2)}/a_{22}^{(2)} & 1 & & \\ & \vdots & & \ddots & \\ & -a_{n2}^{(2)}/a_{22}^{(2)} & & & 1 \end{bmatrix}.$$

We write

$$\mathbf{L}_2 \mathbf{L}_1 \mathbf{A} = [a_{ij}^{(3)}].$$

We carry on in this way and at the end of the s-th stage in the Gaussian triangularization we have obtained

$$\mathbf{L}_s \dots \mathbf{L}_2 \mathbf{L}_1 \mathbf{A} = \left[\begin{array}{cccc|c} a_{11}^{(1)} & & & & \\ & a_{22}^{(2)} & & & \\ & & \ddots & & \\ 0 & & & a_{ss}^{(s)} & \\ \hline & & 0 & & \\ \end{array}\right] \begin{array}{l} \updownarrow s \\ \\ \updownarrow n-s \end{array}$$

where the elements in the right hand blocks are irrelevant. Here the $\mathbf{L}$'s are certain lower triangular matrices with unit diagonals. Restricting our attention to the leading $s \times s$ block, and taking determinants on both sides we find

$$\det [\mathbf{A}]_s = a_{11}^{(1)} a_{22}^{(2)} \dots a_{ss}^{(s)}$$

since the determinants of the blocks in the $\mathbf{L}$-matrices are all 1. This result being true for each s we have

$$a_{ss}^{(s)} = \det [\mathbf{A}]_s / \det [\mathbf{A}]_{s-1}.$$

This result checks with

$$\det \mathbf{A} = \prod_{s=1}^{n} a_{ss}^{(s)}$$

and makes very clear the role of the condition that all the leading minors of $\mathbf{A}$ should be non-singular.

4.4. Solution

All that it is necessary to establish is that all the leading submatrices of $\mathbf{A}$ are non-singular.

(a) If $\mathbf{A}$ has a strictly dominant diagonal, i.e.,

$$(1) \qquad |a_{ii}| > \sum_{\substack{j=1 \\ j \neq i}}^{n} |a_{ij}| \quad \text{for} \quad i = 1, 2, \dots, n$$

so has every leading submatrix $[\mathbf{A}]_r$. For (1) surely implies that

$$|a_{ii}| > \sum_{\substack{j=1 \\ j \neq i}}^{r} |a_{ij}| \quad \text{for} \quad i = 1, 2, \dots, r.$$

We now appeal to the Dominant Diagonal Theorem (see p. 54) to conclude that every leading submatrix is non-singular.

(b) One of the properties of a positive definite matrix is that all its leading principal minors, i.e., $\det [\mathbf{A}]_r$, are positive, so that $[\mathbf{A}]_r$ is certainly non-singular (see p. 58).

4.5. Solution

We discuss the "square case" leaving the rectangular one to the reader. We consider the representation

$$\mathbf{F} = \mathbf{\Phi} \mathbf{R} \tag{1}$$

and denote by $x_1, \ldots, x_n$ the columns of a matrix $\mathbf{X}$. Writing out (1) in terms of the columns we get

$$f_1 = r_{11} \varphi_1$$

$$f_2 = r_{12} \varphi_1 + r_{22} \varphi_2$$

$$\ldots$$

$$f_n = r_{1n} \varphi_1 + r_{2n} \varphi_2 + \ldots + r_{nn} \varphi_n.$$

We also have

$$\varphi_i' \varphi_j = \delta_{ij}, \quad i, j = 1, 2, \ldots, n.$$

The unknown quantities can be obtained by a recurrence process. We first find $r_{11} = \|f_1\|_2$ and take $\varphi_1 = f_1/r_{11}$. Supposing we have the first s vectors φ_i and the corresponding $r_{ij}\,(i \leqq j \leqq s)$ determined. Then the condition that φ_{s+1} is orthogonal to $\varphi_1, \ldots, \varphi_s$ gives the relations

$$r_{j,s+1} = \varphi_j' f_{s+1}, \quad j = 1, 2, \ldots, s,$$

while the normality of φ_{s+1} determines $r_{s+1,s+1}$ by

$$r_{s+1,s+1} = \pm \|f_{s+1} - r_{1,s+1} \varphi_1 - \ldots - r_{s,s+1} \varphi_s\|_2.$$

Each of these calculations involves about $n\,s$ multiplications and the latter also requires a square root. The total cost is therefore about

$$\sum (2ns) \doteqdot n^3$$

operations.

A second solution to this problem can be obtained by using the $\mathbf{LL'}$ scheme and using results given elsewhere in the text or problems and solutions. The main steps, in the "square case" are:

$$\text{compute } \mathbf{F'F}$$

$$\text{factorize } \mathbf{F'F} = \mathbf{LL'}$$

and then

$$\mathbf{R} = \mathbf{L'}.$$

The first stage requires about $n^3/2$ multiplications and the second about $n^3/6$ — in each case we take account of symmetry.

To compute $\mathbf{\Phi} = \mathbf{FR}^{-1}$ requires additionally the inversion of R — costing about $n^3/6$ multiplications and then the multiplication $\mathbf{FR}^{-1}$ which costs about $n^3/2$ multiplications.

We illustrate this process in the rectangular case by a numerical example.

Suppose we want to orthonormalize the two vectors

$$f_1=[1, 2, 3]', \quad f_2=[4, 5, 6]'.$$

We write

$$\mathbf{F}=\begin{bmatrix}1 & 4\\ 2 & 5\\ 3 & 6\end{bmatrix}$$

and we seek an $\mathbf{L}$ such that $\mathbf{FL}'=\boldsymbol{\Phi}$ is orthogonal, i.e., so that $\boldsymbol{\Phi}'\boldsymbol{\Phi}=\mathbf{I}$. This means that

$$(\mathbf{L}\,\mathbf{F}')(\mathbf{F}\,\mathbf{L}')=\boldsymbol{\Phi}'\boldsymbol{\Phi}=\mathbf{I}$$

and so

$$\mathbf{F}'\mathbf{F}=\mathbf{L}^{-1}\mathbf{L}^{-1'}.$$

We compute

$$\mathbf{F}'\mathbf{F}=\begin{bmatrix}14 & 32\\ 32 & 77\end{bmatrix}.$$

We then apply the LDU theorem to find a, b, c so that

$$\begin{bmatrix}a & 0\\ b & c\end{bmatrix}\begin{bmatrix}a & b\\ 0 & c\end{bmatrix}=\begin{bmatrix}14 & 32\\ 32 & 77\end{bmatrix}.$$

We find $a=\sqrt{14}$, $b=32/\sqrt{14}$ and $c=\sqrt{27/7}$. Thus we may take

$$\mathbf{L}^{-1}=\begin{bmatrix}\sqrt{14} & 0\\ 32/\sqrt{14} & \sqrt{27/7}\end{bmatrix}.$$

This gives

$$\mathbf{L}=\frac{1}{\sqrt{54}}\begin{bmatrix}\sqrt{27/7} & 0\\ -32/\sqrt{14} & \sqrt{14}\end{bmatrix}=\begin{bmatrix}1/\sqrt{14} & 0\\ -16/\sqrt{7\cdot 27} & \sqrt{7/27}\end{bmatrix}$$

and finally

$$\boldsymbol{\Phi}=\mathbf{FL}'=\begin{bmatrix}1 & 4\\ 2 & 5\\ 3 & 6\end{bmatrix}\begin{bmatrix}1/\sqrt{14} & -16/\sqrt{7\cdot 27}\\ 0 & \sqrt{7/27}\end{bmatrix}=\begin{bmatrix}1/\sqrt{14}, & 4/\sqrt{21}\\ 2/\sqrt{14}, & 1/\sqrt{21}\\ 3/\sqrt{14}, & -2/\sqrt{21}\end{bmatrix}$$

and orthonormality is easily checked.

4.6. Solution.

We compute

$$\mathbf{A}'\mathbf{A}=\begin{bmatrix}2 & 1 & \sqrt{3}\\ -4 & 0 & 0\\ -2 & -4 & 0\end{bmatrix}\begin{bmatrix}2 & -4 & -2\\ 1 & 0 & -4\\ \sqrt{3} & 0 & 0\end{bmatrix}=\begin{bmatrix}8 & -8 & -8\\ -8 & 16 & 8\\ -8 & 8 & 20\end{bmatrix}$$

and then factorize $\mathbf{A}'\mathbf{A}=\mathbf{L}\mathbf{L}'$ where

$$\mathbf{L}=\begin{bmatrix} 2\sqrt{2} & 0 & 0 \\ -2\sqrt{2} & 2\sqrt{2} & 0 \\ -2\sqrt{2} & 0 & 2\sqrt{3} \end{bmatrix}.$$

If $\mathbf{A}=\boldsymbol{\Phi}\mathbf{R}$ then $\mathbf{A}'=\mathbf{R}'\boldsymbol{\Phi}'$ and

$$\mathbf{A}'\mathbf{A}=\mathbf{R}'\boldsymbol{\Phi}'\boldsymbol{\Phi}\mathbf{R}=\mathbf{R}'\mathbf{R}$$

so that

$$\mathbf{R}=\mathbf{L}'$$

and

$$\boldsymbol{\Phi}=\mathbf{A}\mathbf{L}'^{-1}.$$

We find

$$\mathbf{L}'^{-1}=\begin{bmatrix} \frac{1}{2\sqrt{2}} & \frac{1}{2\sqrt{2}} & \frac{1}{2\sqrt{3}} \\ 0 & \frac{1}{2\sqrt{2}} & 0 \\ 0 & 0 & \frac{1}{2\sqrt{3}} \end{bmatrix}$$

and

$$\boldsymbol{\Phi}=\begin{bmatrix} \frac{1}{\sqrt{2}} & -\frac{1}{\sqrt{2}} & 0 \\ \frac{1}{2\sqrt{2}} & \frac{1}{2\sqrt{2}} & \frac{-\sqrt{3}}{2} \\ \frac{\sqrt{3}}{2\sqrt{2}} & \frac{\sqrt{3}}{2\sqrt{2}} & \frac{1}{2} \end{bmatrix}.$$

Orthogonality of $\boldsymbol{\Phi}$ is easily checked.

(2)
$$\mathbf{B}=\frac{1}{9}\begin{bmatrix} -7 & -4 & -4 \\ 4 & 1 & -8 \\ 4 & -8 & 1 \end{bmatrix}\begin{bmatrix} 1 & 2 & 3 \\ 0 & 4 & 5 \\ 0 & 0 & 6 \end{bmatrix}.$$

4.7. Solution

This is a matter of elementary algebra. The result shows that two 2×2 matrices (even *block* matrices) can be multiplied at the expense of 7 multiplications and 18 additions as compared to the 8 multiplications and 4 additions required in the obvious way.

Building on this result V. Strassen (Numer. Math., *13*, 354—356 (1969)) has shown that two $n\times n$ matrices can be multiplied in $\mathcal{O}(n^{\log_2 7})$ operations;

the same is true for our basic problems: $\mathbf{A}^{-1}$, det $\mathbf{A}$, $\mathbf{A}\boldsymbol{x}=\boldsymbol{b}$. These schemes *are* practical and have been implemented by R. P. Brent (Numer. Math., *16*, 145—156 (1970)).

Another fast matrix multiplication technique has been developed by S. Winograd (Comm. Pure Appl. Math., *23*, 165—179 (1970)).

4.8. Solution

This is a matter of elementary algebra. The results show, for instance, that the inversion of a $2n\times 2n$ matrix can be accomplished by operations on $n\times n$ matrices, a fact which can extend the range of n for which all the work can be done internally.

It is easy to verify that $\mathbf{A}_{11}^{-1}=\mathbf{X}-\mathbf{Y}\mathbf{W}^{-1}\mathbf{Z}$.

4.9. Solution

This is a matter of elementary algebra. The result shows that the inversion of complex matrices can be accomplished by real operations. This form of the solution is due to K. Holladay.

4.10. Solution

This can be regarded as the result of an LU (or LDU) decomposition of a block matrix:

$$\mathcal{A}=\begin{bmatrix}\mathbf{A} & \mathbf{B}\\ \mathbf{C} & \mathbf{D}\end{bmatrix}=\begin{bmatrix}\mathbf{A} & 0\\ \mathbf{C} & \mathbf{I}\end{bmatrix}\begin{bmatrix}\mathbf{I} & \mathbf{A}^{-1}\mathbf{B}\\ 0 & \mathbf{D}-\mathbf{C}\mathbf{A}^{-1}\mathbf{B}\end{bmatrix}$$

or

$$\mathcal{A}=\begin{bmatrix}\mathbf{I} & 0\\ \mathbf{C}\mathbf{A}^{-1} & \mathbf{I}\end{bmatrix}\begin{bmatrix}\mathbf{A} & 0\\ 0 & \mathbf{D}-\mathbf{C}\mathbf{A}^{-1}\mathbf{B}\end{bmatrix}\begin{bmatrix}\mathbf{I} & \mathbf{A}^{-1}\mathbf{B}\\ 0 & \mathbf{I}\end{bmatrix}$$

which can be verified by multiplication. Taking determinants

$$\begin{aligned}\det\mathcal{A}&=\det\begin{bmatrix}\mathbf{A} & 0\\ \mathbf{C} & \mathbf{I}\end{bmatrix}\det\begin{bmatrix}\mathbf{I} & \mathbf{A}^{-1}\mathbf{B}\\ 0 & \mathbf{D}-\mathbf{C}\mathbf{A}^{-1}\mathbf{B}\end{bmatrix}\\ &=\det\mathbf{A}\det(\mathbf{D}-\mathbf{C}\mathbf{A}^{-1}\mathbf{B})\\ &=\det(\mathbf{A}\mathbf{D}-\mathbf{A}\mathbf{C}^{-1}\mathbf{B}).\end{aligned}$$

The matrix $\mathbf{D}-\mathbf{C}\mathbf{A}^{-1}\mathbf{B}$ is called the Schur complement of $\mathbf{A}$ in $\mathcal{A}$.

If $\mathbf{D}$ is non-singular we obtain in the same way.

$$\begin{bmatrix}\mathbf{A} & \mathbf{B}\\ \mathbf{C} & \mathbf{D}\end{bmatrix}=\begin{bmatrix}\mathbf{A}-\mathbf{B}\mathbf{D}^{-1}\mathbf{C} & \mathbf{B}\mathbf{D}^{-1}\\ 0 & \mathbf{I}\end{bmatrix}\begin{bmatrix}\mathbf{I} & 0\\ \mathbf{C} & \mathbf{D}\end{bmatrix}$$

and hence

$$\det\mathcal{A}=\det\mathbf{D}\det(\mathbf{A}-\mathbf{B}\mathbf{D}^{-1}\mathbf{C}).$$

4.11. Solution

We factorize $\mathbf{A}=\mathbf{L}\mathbf{L}'$ and find that all elements are real. Hence $\mathbf{A}$ is positive definite.

Actually

$$\mathbf{L}=\begin{bmatrix} 9 & 0 & 0 & 0 \\ -4 & 10 & 0 & 0 \\ 3 & -5 & 8 & 0 \\ -2 & 6 & -1 & 7 \end{bmatrix}.$$

To solve the system $\mathbf{A}\boldsymbol{x}=\boldsymbol{b}$, i.e., $\mathbf{L}\mathbf{L}'\boldsymbol{x}=\boldsymbol{b}$ we find first

$$\mathbf{L}^{-1}=\begin{bmatrix} 1/9 & 0 & 0 & 0 \\ 2/45 & 1/10 & 0 & 0 \\ -1/72 & 1/16 & 1/8 & 0 \\ -1/120 & -43/560 & 1/56 & 1/7 \end{bmatrix}.$$

We have

$$\boldsymbol{x}=(\mathbf{L}\mathbf{L}')^{-1}\boldsymbol{b}=(\mathbf{L}')^{-1}\mathbf{L}^{-1}\boldsymbol{b}=(\mathbf{L}^{-1})'\cdot\mathbf{L}^{-1}\boldsymbol{b}.$$

We find

$$(\mathbf{L}^{-1}\boldsymbol{b})'=[28,\ 26,\ 15,\ 7]'$$

and then

$$\boldsymbol{x}=(\mathbf{L}^{-1})'[28,\ 26,\ 15,\ 7]'=[4,\ 3,\ 2,\ 1]'.$$

4.12. Solution

No. If we proceed to find the $\mathbf{L}\mathbf{L}'$ decomposition we get:

$$l_{11}^2=1 \text{ so that } l_{11}=1 \text{ say}$$

$$l_{11}l_{12}=2 \text{ so that } l_{12}=2$$

$$l_{12}^2+l_{22}^2=3 \text{ so that } l_{22}^2=-1$$

which demonstrates that the matrix is not positive definite.

Carrying on we find

$$\mathbf{L}=\begin{bmatrix} 1 & . & . & . \\ 2 & i & . & . \\ 3 & 2i & i & . \\ 4 & 3i & i & 1 \end{bmatrix},\quad \mathbf{L}^{-1}=\begin{bmatrix} 1 & . & . & . \\ 2i & -i & . & . \\ -i & 2i & -i & . \\ 1 & -1 & -1 & 1 \end{bmatrix},\quad \mathbf{A}^{-1}=\begin{bmatrix} -3 & 3 & -2 & 1 \\ 3 & -4 & 3 & -1 \\ -2 & 3 & 0 & 1 \\ 1 & -1 & -1 & 1 \end{bmatrix}.$$

If $\mathbf{A}\boldsymbol{x}=[30, 40, 43, 57]'$ we find $\boldsymbol{x}=[1, 2, 3, 4]'$.

4.13. Solution

Premultiplication by a diagonal matrix means multiplication of the rows by the corresponding diagonal element:

$$[\mathbf{D}^{-1}\mathbf{A}]_{ij}=d_i^{-1}a_{ij}$$

and postmultiplication means multiplication of the columns by the corresponding diagonal element:

$$[\mathbf{D}^{-1}\mathbf{A}\mathbf{D}]_{ij}=d_i^{-1}a_{ij}d_j.$$

In order that $\mathbf{D}^{-1}\mathbf{AD}$ should be symmetric, we must have, for all $i, j, i \neq j$

$$d_i^{-1} a_{ij} d_j = d_j^{-1} a_{ji} d_i$$

so that

$$\frac{d_i}{d_j} = \sqrt{\frac{a_{ij}}{a_{ji}}}.$$

We can take $d_1 = 1$ and then

$$d_{j+1} = d_j \sqrt{\frac{a_{j+1,j}}{a_{j,j+1}}} \quad j = 1, 2, \dots, n-1.$$

We assume that $a_{i,j} \neq 0$ for $|i-j| = 1$. The value of the elements in the $(i, i+1)$ and $(i+1, i)$ place is clearly

$$\sqrt{d_{i,i+1} d_{i+1,i}}.$$

4.14. Solution

For the results see Problem 5.13 (iv), (v).

4.15. Solution

In order to check your results use Problem 5.13 (iv), (v).

4.16. Solution

Consider the determination of the r-th row of $\mathbf{L}^{-1} = \mathbf{X}$ from $\mathbf{XL} = \mathbf{I}$. The system of equations to be satisfied is

$$\begin{aligned} x_{r1} l_{11} + x_{r2} l_{21} + \dots + x_{rr} l_{r1} &= 0 \\ x_{r1} 0 + x_{r2} l_{22} + \dots + x_{rr} l_{r2} &= 0 \\ &\vdots \\ x_{rr} l_{rr} &= 1. \end{aligned}$$

This is a triangular system. We have seen that it can be solved by back-substitution at the cost of about $r^2/2$ multiplications. In all we need

$$\sum_{r=1}^{n} r^2/2 \doteqdot n^3/6.$$

See diagram.

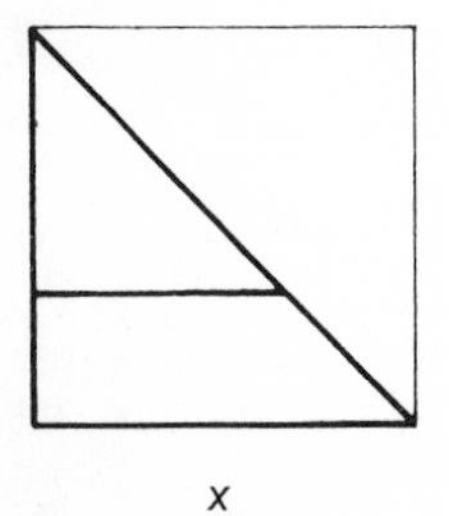

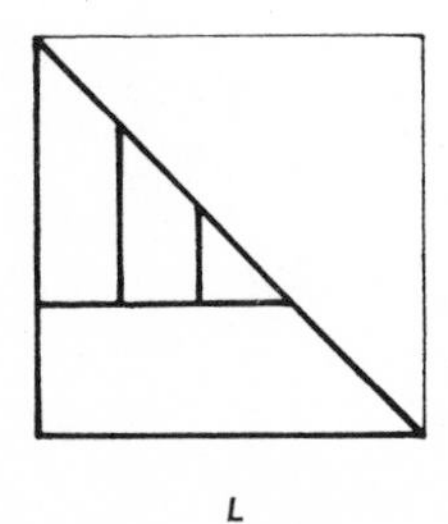

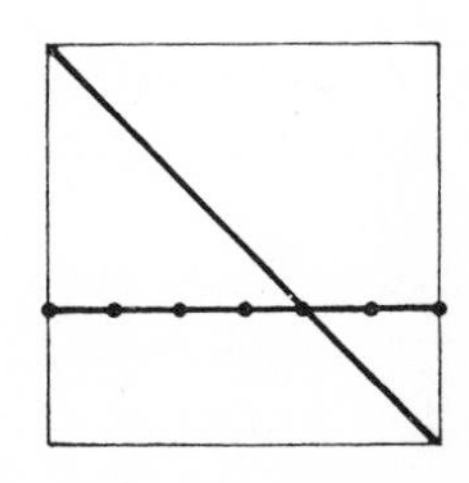

X L = I

4.17. Solution

This is left to the reader. Apparently this problem does not occur frequently in computing practice, although the solution of $\mathbf{X}\boldsymbol{x}=\boldsymbol{b}$ and the evaluation of det $\mathbf{X}$ in this case are of great importance.

4.18. Solution

The Dekker Matrix

This is a matrix $\mathbf{M}$ derived from $\mathbf{H}_n$ which has the advantage, as a test matrix, that both it and its inverse have integral elements. In fact $\mathbf{M}$ is a positive matrix and $\mathbf{M}^{-1}$ is got by changing the sign of alternate elements in $\mathbf{M}$, specifically, $[\mathbf{M}^{-1}]_{ij}=(-1)^{i+j}[\mathbf{M}]_{ij}$.

We have recorded the inverse of the Hilbert matrix $\mathbf{H}_n$ (Problem 5.13 (viii))

$$[\mathbf{H}_n^{-1}]_{ij}=(-1)^{i+j}f_i f_j/(i+j-1)$$

where

$$f_i=(n+i-1)!/(((i-1)!)^2(n-i)!).$$

It is easy to check that f_i is an integer: the quotient $(n+1-i)!/((n-i)!)$ is a product of $2i-1$ consecutive integers, and we use twice the well-known fact that the product of any r consecutive integers is divisible by $r!$ — it is, indeed, a binomial coefficient.

If we write $\mathbf{F}=\text{diag}[f_1, f_2, \ldots, f_n]$, $\mathbf{E}=\text{diag}[-1, 1, -1, \ldots, (-1)^n]$ we have

$$\mathbf{H}^{-1}=\mathbf{FEHEF}.$$

We next define $\mathbf{M}=\mathbf{FG}^{-1}\mathbf{HG}$ where $\mathbf{G}=\text{diag}[g_1, g_2, \ldots, g_n]$ is any non-singular diagonal matrix and observe that

$$\begin{aligned}\mathbf{M}^{-1}&=\mathbf{G}^{-1}\mathbf{H}^{-1}\mathbf{G}\mathbf{F}^{-1}\\ &=\mathbf{G}^{-1}(\mathbf{FEHEF})\mathbf{G}\mathbf{F}^{-1}\\ &=\mathbf{G}^{-1}\mathbf{FEHE}(\mathbf{F}\mathbf{G}\mathbf{F}^{-1})\\ &=(\mathbf{G}^{-1}\mathbf{FE})\mathbf{H}(\mathbf{EG})\\ &=\mathbf{E}(\mathbf{FG}^{-1}\mathbf{HG})\mathbf{E}\\ &=\mathbf{EME}\end{aligned}$$

where we use repeatedly the fact that diagonal matrices commute. This shows that the inverse of $\mathbf{M}$ is got by multiplying its i, j element by $(-1)^{i+j}$ for each i, j.

We shall now show that if we choose $\mathbf{G}$ properly we can ensure that the elements of $\mathbf{M}$ are integers:

If an integer N is expressed in prime factors $N=p_1^{n_1}p_2^{n_2}\ldots p_r^{n_r}$ we define $\tilde{N}=p_1^{m_1}p_2^{m_2}\ldots p_r^{m_r}$ where $m_s=[n_s/2]$, the integral part of $n_s/2$ so that n_s is either

$2m_s$ or $2m_s+1$. We now define

$$g_s=\tilde{f}_s.$$

We begin by showing that $(i+j-1)^2$ is a factor of $f_i f_j$. It is clear that $f_i f_j$ is the product of two sets of $i+j-1$ consecutive numbers $n-j+1, \ldots, n+i-1$ and $(n-i+1), \ldots, (n+j-1)$ divided by $[(i-1)!(j-1)!]^2$. Such a product is divisible by $(i+j-1)!$ and therefore by $(i+j-1)$, and by $(i-1)!$ and the product of the $(j-1)$ remaining consecutive numbers and, a fortiori, by $(j-1)!$. This establishes our assertion, and, incidentally, the fact that $\mathbf{H}_n^{-1}$ has integral elements.

The i, j element of $\mathbf{M}$ is $f_i g_i^{-1} g_j/(i+j-1)$. Let p be any prime and let q_i, q_j, q be the powers of p which occur in $f_i, f_j, i+j-1$. Since $(i+j-1)^2$ divides $f_i f_j$ it follows that $q_i+q_j \geqq 2q$. The power of p which occurs in $f_i g_j$ is $q_i+[q_j/2]$ while that which occurs in $g_i(i+j-1)$ is $[q_i/2]+q$ which is not greater than $q_i+[q_j/2]$. To see this write $E=q_i+[q_j/2]-[q_i/2]-q$. Then if q_i and q_j are both even or both odd

$$E=\tfrac{1}{2}q_i+\tfrac{1}{2}q_j-q=\tfrac{1}{2}(q_i+q_j-2q)\geqq 0.$$

If q_i is odd and q_j even

$$E=\tfrac{1}{2}q_i+\tfrac{1}{2}+\tfrac{1}{2}q_j-q\geqq 0.$$

If q_i even and q_j odd then $q_i+q_j\geqq 2q+1$ and

$$E=\tfrac{1}{2}q_i+\tfrac{1}{2}q_j-\tfrac{1}{2}-q\geqq 0.$$

It follows that $[\mathbf{M}]_{ij}$ is integral as asserted.

4.19. Solution

The method for the general case is clear from the following discussion of the example:

Subtract 2 col_2 from col_1 to get a zero in the (1, 1) place. Add 7 col_3 to col_1 to get a zero in the (2, 1) place. Subtract 12 col_4 from col_1 to get a zero in the (3, 1) place. Add 16 col_5 to col_1 to get a zero in the (4, 1) place. We have therefore

$$\det \mathbf{A}=\begin{vmatrix} 0 & 1 & 0 & 0 & 0 \\ 0 & 2 & 1 & 0 & 0 \\ 0 & 3 & 2 & 1 & 0 \\ 0 & 4 & 3 & 2 & 1 \\ 20 & 5 & 4 & 3 & 2 \end{vmatrix}$$

and so

$$\det \mathbf{A}=20\times 1\times 1\times 1\times 1=20.$$

4.20. Solution

This is a rather difficult problem. For accounts of it see, e.g.,
W. Trench, J. *SIAM*, *12*, 515—522 (1964) and *13*, 1102—1107 (1965).
E. H. Bareiss, Numer. Math., *13*, 404—426 (1969).
J. L. Phillips, Math. Comp., *25*, 599—602 (1971).
J. Rissanen, Math. Comp. *25*, 147—154 (1971), and Numer. Math., *22*, 361—366 (1974), S. Zohar, J. ACM, *21*, 272—276 (1974).

4.21. Solution

Suppose

$$\mathbf{U}=\begin{bmatrix} a & b & c & d \\ \cdot & e & f & g \\ \cdot & \cdot & h & i \\ \cdot & \cdot & \cdot & j \end{bmatrix}.$$

Then we have $a^2=1$, $a=1$ say; $ab=\frac{1}{2}$, $b=\frac{1}{2}$; $c=\frac{1}{3}$; $d=\frac{1}{4}$; $b^2+e^2=1$, $e^2=\frac{3}{4}$, $e=\sqrt{3/2}$ say; $bc+ef=\frac{2}{3}$, $f=1/\sqrt{3}$; $g=\sqrt{3/4}$; $c^2+f^2+h^2=1$, $h^2=1-(1/3)-(1/9)$, $h=\sqrt{5}/3$ say; $i=\sqrt{5}/4$, $d^2+g^2+i^2+j^2=1$, $j^2=1-(5/16)-(3/16)-(1/16)$, $j=\sqrt{7}/4$ say. We find easily that

$$\mathbf{U}^{-1}=\begin{bmatrix} 1 & -1/\sqrt{3} & 0 & 0 \\ 0 & 2/\sqrt{3} & -2/\sqrt{5} & 0 \\ 0 & 0 & 3/\sqrt{5} & -3/\sqrt{7} \\ 0 & 0 & 0 & 4/\sqrt{7} \end{bmatrix}$$

and

$$\mathbf{A}^{-1}=\begin{bmatrix} 4/3 & -2/3 & 0 & 0 \\ -2/3 & 32/15 & -6/5 & 0 \\ 0 & -6/5 & 108/35 & -12/7 \\ 0 & 0 & -12/7 & 16/7 \end{bmatrix}.$$

See Problem 5.13 (iii) for the general case.

4.22. Solution

Consider the equations

$$(1) \qquad l_{11}u_{11}=a_{11}, \quad l_{11}u_{12}=a_{12}, \ldots, l_{11}u_{1n}=a_{1n}.$$

Choosing u_{11} in the first determines l_{11} and then $u_{12}, \ldots, u_{1n}$ are determined. So we get the first row of $\mathbf{U}$.

Consider the equations

$$(2)\qquad (l_{11}u_{11}=a_{11}),\quad l_{21}u_{11}=a_{21},\ \dots,\ l_{n1}u_{11}=a_{n1}$$

and we see that the first column of **L** is determined by the value of u_{11}.

Now suppose we have determined the first $r-1$ rows of **U** and the first $r-1$ columns of **L**. We write down the equations corresponding to (1), (2) above:

$$(3_i)\qquad a_{ri}=\sum_{k=1}^{r} l_{rk}u_{ki}=\sum_{k=1}^{r-1} l_{rk}u_{ki}+l_{rr}u_{ri}\quad i=r, r+1, \dots, n.$$

$$(4_i)\qquad a_{ir}=\sum_{k=1}^{r-1} l_{ik}u_{kr}+l_{ir}u_{rr}\quad i=(r), r+1, \dots, n.$$

From (3_r), by choice of u_{rr} we determine l_{rr} and then from $(3_{r+1}), (3_{r+2}), \dots, (3_n)$ we determine $u_{r,r+1}, \dots, u_{r,n}$, using the available l's and u's. In the same way, from $(4_{r+1}), (4_{r+2}), \dots, (4_n)$ we determine $l_{r+1,r}, \dots, l_{nr}$. Thus the r's and the l's can be computed by rows and columns.

Let us count the multiplications and divisions involved in the r-th stage. We require to handle

$(n-r+1)$ equations each involving $(r-1)$ products and one division, i.e. $(n-r+1)r$ operations and

$(n-r)$ equations each involving $(r-1)$ products and one division, i.e. $(n-r)r$ operations. That is, altogether, about

$$2\sum_{r=1}^{n}(n-r)r=2n\sum r-2\sum r^2 \doteqdot n^3-\frac{2n^3}{3}=\frac{1}{3}n^3.$$

Consider next the product $\mathbf{UL}=\mathbf{A}$. Let us look at the computation of the r-th row of the product. Because of the zeros at the beginning the computation of each of

$$a_{r1}, a_{r2}, \dots, a_{rr}$$

will involve $n-r$ multiplications, in all, $r(n-r)$. The computation of

$$a_{r,r+1}, a_{r,r+2}, \dots, a_{r,n}$$

requires

$$(n-r-1)+(n-r-2)+\dots+1=\frac{(n-r)(n-r-1)}{2}$$

multiplications. Together we require

$$\sum\left\{r(n-r)+\frac{(n-r)(n-r-1)}{2}\right\}\doteqdot\frac{n^3}{6}.$$

From Problem 4.16, the determination of $\mathbf{L}^{-1}$, $\mathbf{U}^{-1}$ each involves about $n^3/6$ multiplications.

The grand total is therefore $\dfrac{n^3}{3}+\dfrac{n^3}{3}+2\dfrac{n^3}{6}=n^3$.

See diagrams.

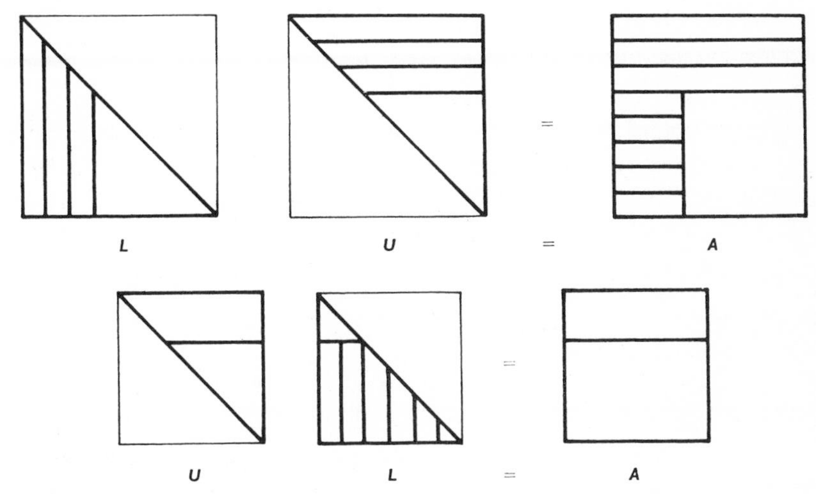

4.23. Solution

Use Problem 4.10 above. See SIAM Review *14*, 499—500 (1972).

4.24. Solution

Since $1=\mathbf{x}^*\mathbf{x}=x_1^2+\hat{\mathbf{x}}^*\hat{\mathbf{x}}$ we have $\hat{\mathbf{x}}^*\hat{\mathbf{x}}=1-x_1^2$. It is clear that $\mathbf{U}$ is hermitian and we compute

$$\mathbf{U}^*\mathbf{U}=\mathbf{U}^2=$$

$$=\begin{bmatrix} x_1^2+\hat{\mathbf{x}}^*\hat{\mathbf{x}} & x_1\hat{\mathbf{x}}^*-\hat{\mathbf{x}}^*+(1+x_1)^{-1}\hat{\mathbf{x}}^*\hat{\mathbf{x}}\hat{\mathbf{x}}^* \\ x_1\hat{\mathbf{x}}^*-\hat{\mathbf{x}}^*+(1+x_1)^{-1}\hat{\mathbf{x}}\hat{\mathbf{x}}^*\hat{\mathbf{x}} & \hat{\mathbf{x}}\hat{\mathbf{x}}^*+\hat{\mathbf{I}}-2(1+x_1)^{-1}\hat{\mathbf{x}}\hat{\mathbf{x}}^*+(1+x_1)^{-2}\hat{\mathbf{x}}\hat{\mathbf{x}}^*\hat{\mathbf{x}}\hat{\mathbf{x}}^* \end{bmatrix}.$$

The (1,1) element is clearly 1. In the case of the (1,2) element we can replace the third term by $(1+x_1)^{-1}(\hat{\mathbf{x}}^*\hat{\mathbf{x}})\hat{\mathbf{x}}^*=(1+x_1)^{-1}(1-x_1^2)\hat{\mathbf{x}}^*=(1-x_1)\hat{\mathbf{x}}^*$ and this cancels out the first two terms; similarly for the (2,1) element. The (2,2) element can be written as

$$\begin{aligned} &\hat{\mathbf{I}}+\hat{\mathbf{x}}\hat{\mathbf{x}}^*(1-2(1+x_1)^{-1})+(1+x_1)^{-2}\hat{\mathbf{x}}(\hat{\mathbf{x}}^*\hat{\mathbf{x}})\hat{\mathbf{x}}^* \\ &=\hat{\mathbf{I}}+\hat{\mathbf{x}}\hat{\mathbf{x}}^*((x_1-1)(1+x_1)^{-1})+(1+x_1)^{-2}(1-x_1^2)\hat{\mathbf{x}}\hat{\mathbf{x}}^* \\ &=\hat{\mathbf{I}}+\hat{\mathbf{x}}\hat{\mathbf{x}}^*[(x_1-1)(1+x_1)^{-1}+(1+x_1)^{-1}(1-x_1)] \\ &=\hat{\mathbf{I}}. \end{aligned}$$

Hence $\mathbf{U}^*\mathbf{U}=\mathbf{I}$.

4.25. Solution

The standard reduction processes, which are variants of the Gaussian elimination process, are effective in the context of *theoretical arithmetic*. However we clearly cannot handle such simple cases as

$$\begin{bmatrix} 3 & 4 \\ 4 & 5 \end{bmatrix} \quad \text{or} \quad \begin{bmatrix} 1 & 2 & 3 \\ 2 & 7 & 8 \\ 3 & 13 & 10 \end{bmatrix}$$

exactly on a decimal machine. It is, however, possible to handle this problem exactly when the coefficients are rational provided we have a "rational arithmetic package" incorporated in our machine. This must include subroutines for addition and multiplication of rationals (regarded as ordered pairs of integers)

$$(n_1, d_1) + (n_2, d_2) = (n_1 d_2 + n_2 d_1, d_1 d_2)$$

$$(n_1, d_1) \times (n_2, d_2) = (n_1 n_2, d_1 d_2).$$

In addition, there should be a subroutine for "reduction to lowest terms", so as to keep the integer components as small as possible: i.e., we want a Euclidean algorithm to obtain $d = gcd(n_1, d_1)$ and then replace (n_1, d_1) by $(n_1/d, d_1/d)$.

4.26. Solution

Similar remarks apply here as in Problem 4.25. Indeed all questions about rank and linear dependence and independence, while appropriate in theoretical arithmetic (and, in practice, when we have rational arithmetic available), are largely meaningless in the context of practical computation. For instance, arbitrarily small perturbations of the zero matrix can manifestly produce matrices of any rank.

4.27. Solution

$$\mathcal{A}^{-1} = \begin{bmatrix} \mathbf{I} & -\mathbf{A} & \mathbf{AB} \\ 0 & \mathbf{I} & -\mathbf{B} \\ 0 & 0 & \mathbf{I} \end{bmatrix}.$$

This shows that the product $\mathbf{AB}$ can be obtained by the inversion of the $3n \times 3n$ matrix $\mathcal{A}$.

4.28. Solution

It is easy to verify that

$$\det(\mathbf{A} - \lambda \mathbf{I}) = (1 + \lambda)(1 + 9\lambda - \lambda^2)$$

so that -1 is a characteristic value and so $\mathbf{A}$ cannot be positive definite. Alternatively, the quadratic form

$$x^2+3y^2+4z^2+4xy+6xz+8yz$$

has the value -5 for $x=2$, $y=1$, $z=-2$.

Although $\det[1]=1$, $\det \mathbf{A}=1$, we have $\det\begin{bmatrix}1 & 2\\ 2 & 3\end{bmatrix}=-1$.

Chapter 5

5.1. Solution

$$\det \mathbf{A}=1, \quad \det \mathbf{B}=-118.94, \quad \det \mathbf{C}_4=5, \quad \det(\mathbf{C}_4+\mathbf{E})=4.96.$$

$$\mathbf{A}^{-1}=\begin{bmatrix}-265, & 108, & 366\\ -2920, & 1190, & 4033\\ 8684 & -3539, & -11994\end{bmatrix}, \quad \mathbf{C}_4^{-1}=-\frac{1}{5}\begin{bmatrix}4 & 3 & 2 & 1\\ 3 & 6 & 4 & 2\\ 2 & 4 & 6 & 3\\ 1 & 2 & 3 & 4\end{bmatrix}.$$

The first result can be simply obtained from the fact that, since $\det \mathbf{A}=1$,

$$[\mathbf{A}^{-1}]_{ij}=\text{cofactor of } ji \text{ element in } \mathbf{A}.$$

The second result can be derived by obtaining (as in Problem 5.4 below) the factorization

$$-\mathbf{C}_4=\mathbf{L}\mathbf{L}' \quad \text{where} \quad \mathbf{L}=\begin{bmatrix}\sqrt{2} & 0 & 0 & 0\\ -1/\sqrt{2} & \sqrt{3/2} & 0 & 0\\ 0 & -\sqrt{2/3} & \sqrt{4/3} & 0\\ 0 & 0 & -\sqrt{3/4} & \sqrt{5/4}\end{bmatrix}=\mathbf{L}_1\mathbf{D}^{1/2}$$

where

$$\mathbf{L}_1=\begin{bmatrix}1 & 0 & 0 & 0\\ -\frac{1}{2} & 1 & 0 & 0\\ 0 & -\frac{2}{3} & 1 & 0\\ 0 & 0 & -\frac{3}{4} & 1\end{bmatrix}, \quad \mathbf{D}=\operatorname{diag}\left[2, \frac{3}{2}, \frac{4}{3}, \frac{5}{4}\right].$$

Then we have

$$\mathbf{C}_4^{-1}=\mathbf{L}_1^{-1\prime}\mathbf{D}^{-1}\mathbf{L}_1^{-1}$$

where

$$\mathbf{L}_1^{-1}=\begin{bmatrix}1 & 0 & 0 & 0\\ \frac{1}{2} & 1 & 0 & 0\\ \frac{1}{3} & \frac{2}{3} & 1 & 0\\ \frac{1}{4} & \frac{2}{4} & \frac{3}{4} & 1\end{bmatrix}.$$

$$\|\mathbf{A}\|_\infty=183, \quad \|\mathbf{A}^{-1}\|_\infty=24217.$$

$$\|\mathbf{A}\|_1=245, \quad \|\mathbf{A}^{-1}\|_1=16393.$$

The characteristic values of

$$\mathbf{A}\mathbf{A}'=\begin{bmatrix}11989 & -968 & 8966\\ -968 & 13445 & -4668\\ 8966 & -4668 & 7869\end{bmatrix}$$

are approximately

$$2.10\times 10^4,\quad 1.23\times 10^4,\quad 3.87\times 10^{-9}$$

so that approximately,

$$\|\mathbf{A}\|_2=145,\quad \|\mathbf{A}^{-1}\|_2=16\,069.$$

$$\|\mathbf{C}_4\|_\infty=\|\mathbf{C}_4\|_1=4,\quad \|\mathbf{C}_4^{-1}\|_\infty=\|\mathbf{C}_4^{-1}\|_1=3.$$

Since $\mathbf{C}_4$ is symmetric we need only compute the characteristic values of $\mathbf{C}_4$. These are $-2+2\cos(k\pi/5)=-4\sin^2 k\pi/10$, $k=1, 2, 3, 4$. Hence

$$\|\mathbf{C}_4\|_2=4\sin^2\frac{4\pi}{10}=4\cos^2\frac{\pi}{10},\quad \|\mathbf{C}_4^{-1}\|_2=4\sin^2\frac{\pi}{10}.$$

The characteristic polynomial of $\mathbf{A}$ is $-\lambda^3+3\lambda^2+11069\lambda+1$ and this has zeros 106.7, $-0.00009\,034$, -103.7 approximately.

5.2. Solution

(W. Kahan, Canadian Math. Bull., *9,* 757—801 (1966).)

(a)
$$\mathbf{x}=\begin{bmatrix}2\\-2\end{bmatrix}$$

$$\mathbf{r}=\begin{bmatrix}-10^{-8}\\ 10^{-8}\end{bmatrix}.$$

$$\mathbf{A}^{-1}=-10^8\begin{bmatrix}.8648 & -.1441\\ -1.2969 & .2161\end{bmatrix}.$$

$$\|\mathbf{A}\|_\infty=2.1617,\quad \|\mathbf{A}^{-1}\|_\infty=1.513\times 10^8.$$

$$\|\mathbf{A}\|_1=1.513,\quad \|\mathbf{A}^{-1}\|_1=2.1617\times 10^8.$$

5.3. Solution

It is obvious that the solution of $\mathbf{W}\mathbf{x}=\mathbf{b}$ in the first case is $\mathbf{x}=\mathbf{e}$.

We deal with a general version of the second case. If

$$\mathbf{W}\mathbf{x}=\mathbf{b}+\boldsymbol{\beta}$$

then

$$\mathbf{x}=\mathbf{W}^{-1}(\mathbf{b}+\boldsymbol{\beta})=\mathbf{W}^{-1}\mathbf{b}+\mathbf{W}^{-1}\boldsymbol{\beta}.$$

Using the result of the first part we have, if $\mathbf{b}'=[32, 22, 33, 31]$,

$$\mathbf{x}=\mathbf{e}+\mathbf{W}^{-1}\boldsymbol{\beta}$$

so that when $\boldsymbol{\beta}'=[\varepsilon, -\varepsilon, \varepsilon, -\varepsilon]$, using Problem 5.4,

$$x_1=1+82\,\varepsilon, \quad x_2=1-136\,\varepsilon, \quad x_3=1+35\,\varepsilon, \quad x_4=1-21\,\varepsilon.$$

5.4. Solution

If we write

$$\mathbf{L}=\begin{bmatrix} a & . & . & . \\ b & c & . & . \\ d & e & f & . \\ g & h & i & j \end{bmatrix}, \quad \mathbf{L}^{-1}=\begin{bmatrix} A & . & . & . \\ B & C & . & . \\ D & E & F & . \\ G & H & I & J \end{bmatrix}$$

we find, equating elements in $\mathbf{L}\mathbf{L}'=\mathbf{W}$

$$a^2=10, \quad a=\sqrt{10},$$

$$ab=7, \quad b=7/\sqrt{10},$$

$$ad=8, \quad d=8/\sqrt{10},$$

$$ag=7, \quad g=7/\sqrt{10};$$

$$b^2+c^2=5, \quad c^2=5-(49/10), \quad c=1/\sqrt{10},$$

$$bd+ce=6, \quad ce=(4/10), \quad e=4/\sqrt{10},$$

$$bg+ch=5, \quad ch=(1/10), \quad h=1/\sqrt{10};$$

$$d^2+e^2+f^2=10, \quad f^2=10-(64/10)-(16/10), \quad f=\sqrt{2},$$

$$dg+eh+fi=9, \quad fi=3, \quad i=3/\sqrt{2};$$

$$g^2+h^2+i^2=j^2=10, \quad j^2=10-(49/10)-(1/10)-(9/2), \quad j=1/\sqrt{2}.$$

In the same way equating elements in $\mathbf{L}\mathbf{L}^{-1}=\mathbf{I}$ we find

$$aA=1; \quad cC=1, \quad bA+cB=0; \quad fF=1, \quad eC+fE=0,$$

$$dA+eB+fD=0; \quad jJ=1, \quad iF+jI=0, \quad hC+iE+jH=0,$$

$$gA+hB+iD+jG=0$$

which gives

$$A=1/\sqrt{10}, \quad C=\sqrt{10}, \quad B=-7/\sqrt{10}, \quad F=1/\sqrt{2}, \quad E=-2\sqrt{2},$$

$$D=\sqrt{2}, \quad J=\sqrt{2}, \quad I=-3/\sqrt{2}, \quad H=5\sqrt{2}, \quad G=-3\sqrt{2}.$$

Multiplying $\mathbf{L}^{-1}(\mathbf{L}^{-1})'$ gives the result required.

Since $\mathbf{W}, \mathbf{W}^{-1}$ are symmetric we have

$$\|\mathbf{W}\|_1=\|\mathbf{W}\|_\infty=33, \quad \|\mathbf{W}^{-1}\|_1=\|\mathbf{W}\|_\infty=136.$$

5.5. Solution

$$\mathbf{L}=\begin{bmatrix}1 & 0 & 0 & 0\\ \frac{1}{2} & 1 & 0 & 0\\ \frac{1}{3} & 1 & 1 & 0\\ \frac{1}{4} & \frac{9}{10} & \frac{3}{2} & 1\end{bmatrix}, \quad D=\operatorname{diag}\left[1, \frac{1}{12}, \frac{1}{180}, \frac{1}{2800}\right].$$

$$\mathcal{L}=\begin{bmatrix}1 & 0 & 0 & 0\\ \frac{1}{2} & 1/2\sqrt{3} & 0 & 0\\ \frac{1}{3} & 1/2\sqrt{3} & 1/6\sqrt{5} & 0\\ \frac{1}{4} & 3\sqrt{3}/20 & 1/4\sqrt{5} & 1/20\sqrt{7}\end{bmatrix}$$

$$\mathbf{H}_4^{-1}=\begin{bmatrix}16 & -120 & 240 & -140\\ -120 & 1200 & -2700 & 1680\\ 240 & -2700 & 6480 & -4200\\ -140 & 1680 & -4200 & 2800\end{bmatrix}$$

$$\|\mathbf{H}_4\|_1=\|\mathbf{H}_4\|_\infty=\frac{25}{12}, \quad \text{(row 1)}$$

$$\|\mathbf{H}_4^{-1}\|_1=\|\mathbf{H}_4^{-1}\|_\infty=13620. \quad \text{(row 3)}$$

We note that

$$\det \mathbf{H}_4=\det \mathbf{D}=\frac{1}{12\times 180\times 2800}=\frac{1}{6048\times 10^3}\doteqdot 1.6534\times 10^{-7}.$$

5.6. Solution

$$\mathbf{A}^{-1}=\begin{bmatrix}1-n^2 & n^2\\ n^2 & -n^2\end{bmatrix}, \quad \mathbf{AB}-\mathbf{I}=\begin{bmatrix}0 & 0\\ n^{-1} & 0\end{bmatrix}, \quad \mathbf{BA}-\mathbf{I}=\begin{bmatrix}n & +n\\ -n & -n\end{bmatrix}.$$

$$\|\mathbf{A}\|_\infty=2, \quad \|\mathbf{A}^{-1}\|_\infty=2n^2, \quad \|\mathbf{B}\|_\infty=2n^2-n$$

$$\|\mathbf{AB}-\mathbf{I}\|_\infty=n^{-1}, \quad \|\mathbf{BA}-\mathbf{I}\|_\infty=2n.$$

5.7. Solution

See F. R. Moulton, Amer. Math. Monthly, *20*, 242—249 (1913).

$$\begin{matrix}-1.8000000\times 10^{-7} & 1.8490000\times 10^{-5} & -1.8150000\times 10^{-5}\\ 2.1000000\times 10^{-7} & 1.7260000\times 10^{-5} & -1.8140000\times 10^{-5}\\ -4.1000000\times 10^{-8} & -1.8749000\times 10^{-5} & 1.8134000\times 10^{-5}.\end{matrix}$$

These were obtained on an IBM 360/50 which gave as solutions

$$x=1.0270566, \quad y=2.0919171, \quad z=-0.38048000.$$

5.8. Solution

See E. H. Neville, Phil. Mag. {7}, *39*, 35—48 (1948).

$$x = 5.38625\ 24221\ 14004\ 89671\ 5,$$
$$y = -2.81334\ 69056\ 56987\ 06591\ 5,$$
$$z = -11.59232\ 35480\ 19317\ 71940\ 9,$$
$$u = 6.36482\ 51116\ 16317\ 76363\ 4,$$
$$v = 7.99287\ 21174\ 39987\ 42297\ 9,$$
$$w = -4.20355\ 33598\ 11286\ 99413\ 8.$$

5.9. Solution

This matrix was introduced by H. Rutishauser (*On test matrices,* pp. 349—365, in: *Programmation en mathématiques numérique*, CNRS, Paris 1968).

It is easily shown that det $\mathbf{R}=1$ and that

$$\mathbf{R}^{-1}=\begin{bmatrix} 105 & 167 & -304 & 255 \\ 167 & 266 & -484 & 406 \\ -304 & -484 & 881 & -739 \\ 255 & 406 & -739 & 620 \end{bmatrix}.$$

The characteristic values of $\mathbf{R}$ are approximately

$$19.1225, \quad 10.88282, \quad 8.99417, \quad .0005343.$$

The condition numbers of $\mathbf{R}$ corresponding to the $2-$ and $\infty-$ vector norms are respectively

$$\varkappa_2(\mathbf{R}) \doteqdot 19.1225/\ .0005343 \doteqdot 35790$$

and

$$\varkappa_\infty(\mathbf{R}) = 26 \times 2408 = 62608.$$

5.10. Solution

In this case it is simplest to find $\mathbf{A}^{-1}$, since we need this for the second part. We have det $\mathbf{A}=330+42+42-99-20-294=1$. Hence since

$$\mathbf{A}^{-1}=(\det \mathbf{A})^{-1}[\hat{\mathbf{A}}_{ji}],$$

where $\hat{\mathbf{A}}_{ji}$ is the ji cofactor in $\mathbf{A}$, we have

$$\mathbf{A}^{-1}=\begin{bmatrix} 62 & -36 & -19 \\ -36 & 21 & 11 \\ -19 & 11 & 6 \end{bmatrix}$$

and so $\boldsymbol{x}=[-36, 21, 11]'$.

Clearly

$$\|\mathbf{A}\|_\infty = 20, \quad \|\mathbf{A}^{-1}\|_\infty = 117, \ \varkappa(\mathbf{A}) = 2340.$$

The characteristic values of $\mathbf{A}$ are 16.662, 5.326 and 0.0112 so that $\varkappa_2(\mathbf{A}) \doteqdot 1487$.

We have

$$\mathbf{C}_3^{-1} = -\frac{1}{4}\begin{bmatrix} 3 & 2 & 1 \\ 2 & 4 & 2 \\ 1 & 2 & 3 \end{bmatrix}$$

so that

$$\|\mathbf{C}_3\|_\infty = 4, \quad \|\mathbf{C}_3^{-1}\|_\infty = 2, \quad \varkappa_\infty(\mathbf{C}_3) = 8.$$

5.11. Solution

We find that the determinant of the system is 1 and that the solution is $x=1$, $y=-3$, $z=-2$.

The inverse of the matrix is

$$\begin{bmatrix} 6 & -4 & -1 \\ -4 & 11 & 7 \\ -1 & 7 & 5 \end{bmatrix}.$$

The condition number with respect to the Chebyshev vector norm is

$$105 \times 22 = 2310$$

which is large for a 3×3 matrix.

5.12. Solution

See F. L. Bauer, ZAMM, *46*, 409—421 (1966).

$$\mathbf{A}^{-1} = \frac{1}{8910}\begin{bmatrix} 10000 & -1100 & 10 \\ -1100 & 11111 & -11 \\ 10 & -11 & 1 \end{bmatrix}$$

$\varkappa_\infty(\mathbf{A}) = 10101 \times (11110/8910) \doteqdot 12596.$

$$\mathbf{D}\,\mathbf{A} = \begin{bmatrix} \frac{1}{3} & \frac{1}{3} & \frac{1}{3} \\ \frac{1}{111} & \frac{10}{111} & \frac{100}{111} \\ \frac{1}{10101} & \frac{100}{10101} & \frac{10000}{10101} \end{bmatrix}, \quad \mathbf{A}^{-1}\mathbf{D}^{-1} = \begin{bmatrix} \frac{1000}{297} & \frac{-4070}{297} & \frac{3367}{297} \\ \frac{-100}{270} & \frac{3737}{270} & \frac{-3367}{270} \\ \frac{10}{2970} & \frac{-407}{2970} & \frac{3367}{2970} \end{bmatrix}.$$

$\varkappa_\infty(\mathbf{D}\,\mathbf{A}) = 1 \times (8437/297) \doteqdot 28.$

This shows how we can improve the condition of a matrix by "scaling", i.e. multiplying by diagonal matrices. In this example the diagonal $\mathbf{D}$ is optimal for one-sided scaling in the Chebyshev norm.

5.13. Solution

(i) $\mathbf{A}_1$ is an orthogonal matrix and so $\mathbf{A}_1^{-1}=\mathbf{A}_1'=\mathbf{A}_1$. This can be established by elementary trigonometry using the relation

$$\sin\alpha+\sin 2\alpha+\ldots+\sin n\alpha=\frac{\sin\left(\frac{1}{2}(n+1)\alpha\right)\sin\frac{1}{2}n\alpha}{\sin\frac{1}{2}\alpha}.$$

We note that $\mathbf{A}_1$ is the modal matrix of $\mathbf{A}_8$, i.e., the columns of $\mathbf{A}_1$ are the characteristic vectors of $\mathbf{A}_8$ so that

$$\mathbf{A}_1\mathbf{A}_8\mathbf{A}_1=\operatorname{diag}[\alpha_1,\ldots,\alpha_n]$$

where $\alpha_1,\ldots,\alpha_n$ are the characteristic values of $\mathbf{A}_8$. See (iv) of this problem and also Problem 6.11.

The characteristic values of $\mathbf{A}$ are necessarily all real, $\mathbf{A}$ being symmetric. Since they have absolute value 1, $\mathbf{A}$ being orthogonal, they must be ± 1. It can be shown that they are

$$1;\ \pm 1;\ \pm 1,\ 1;\ \pm 1,\ \pm 1\ldots$$

for $n=1, 2, 3, 4, \ldots$.

This matrix is *not* positive definite.

(ii) See Problem 1.7.

More generally, the inverse of $\alpha\mathbf{I}+\beta\mathbf{J}$ is of the form $\gamma\mathbf{I}+\delta\mathbf{J}$ because $\mathbf{J}^2=n\mathbf{J}$ and we can solve

$$(\alpha\mathbf{I}+\beta\mathbf{J})(\gamma\mathbf{I}+\delta\mathbf{J})=\mathbf{I}$$

by putting

$$\alpha\gamma=1,\quad \alpha\delta+\beta\gamma+n\beta\delta=0,$$

i.e., if $n\beta+\alpha\neq 0$,

$$\gamma=\alpha,\quad \delta=-\beta/\{\alpha(n\beta+\alpha)\}.$$

In the special case $\alpha=m$, $\beta=1$ we have

$$\gamma=n^{-1},\quad \delta=-1/(2n^2).$$

If we consider $\det(\mathbf{A}_3-\lambda\mathbf{I})$ we can add all the columns to the first and take out a common factor $2n-\lambda$ leaving a matrix with its first row and first column all 1's. If we subtract the first row from each of the others we get a matrix whose determinant is clearly $(n-\lambda)^{n-1}$. Thus the characteristic polynomial is $(2n-\lambda)(n-\lambda)^{n-1}$ so that the characteristic roots are $2n$ and n (with multiplicity $n-1$). The matrix $\mathbf{A}_3$ is positive definite and its determinant is $2n^n$.

(iii) For the case $n=4$ see Problem 4.21.

The more general matrix $\mathbf{A}$ which is symmetric and whose i, j element, for $i\leqq j$, is

$$[\mathbf{A}]_{i,j}=a_i/a_j,$$

where we assume the a's are distinct and non-zero, is discussed in Amer. Math. Monthly, *53*, 534—535 (1946). Specializing to $a_i = i$ we get $\mathbf{A}_7$.

One way of obtaining $\mathbf{A}^{-1}$ is the observation that

$$\text{(1)} \qquad \mathbf{B}\mathbf{A}\mathbf{B}' = \mathbf{D} = \operatorname{diag}[1-b_1^2, 1-b_2^2, \ldots, 1-b_{n-1}^2, 1]$$

where

$$\mathbf{B} = \begin{bmatrix} 1, & -b_1 & & & \\ & 1, & -b_2 & & \\ & & \ddots & & \\ & & & 1, & -b_{n-1} \\ & & & & 1 \end{bmatrix}, \quad b_i = a_i/a_{i+1}.$$

To verify (1), we observe that

$$\text{(2)} \qquad \mathbf{B}\mathbf{A} = \begin{bmatrix} r_1 - b_1 r_2 \\ r_2 - b_2 r_3 \\ \vdots \\ r_{n-1} - b_{n-1} r_n \\ r_n \end{bmatrix}, \quad \mathbf{B}\mathbf{A}\mathbf{B}' = \begin{bmatrix} \gamma_1 - b_1 \gamma_2 \\ \gamma_2 - b_2 \gamma_3 \\ \vdots \\ \gamma_{n-1} - b_{n-1} \gamma_n \\ \gamma_n \end{bmatrix}$$

where the $\boldsymbol{r}$'s are *rows* of $\mathbf{A}$ and the γ's are the *columns* of $\mathbf{BA}$. It is now easy to check that $\mathbf{BA}$ is a lower triangular matrix whose i, j element, $i \geqq j$, is:

$$\frac{a_j}{a_i} - \frac{a_i a_j}{(a_{i+1})^2}$$

where we interpret $a_{n+1} = \infty$, $b_n = 0$ so that the last row is not changed. Similarly we find that $\mathbf{BAB}'$ is the diagonal matrix given.

From (1) we conclude that

$$\mathbf{A}^{-1} = \mathbf{B}' \mathbf{D}^{-1} \mathbf{B}$$

and using the interpretations (2) we find that $\mathbf{A}^{-1}$ is the symmetric triple diagonal matrix given by

$$[\mathbf{A}^{-1}]_{ii} = d_i^{-1} + (a_{i-1}/a_i)^2 d_{i-1}^{-1} \quad (a_0 = 0)$$

$$[\mathbf{A}^{-1}]_{i,i+1} = -a_i/(a_{i+1} d_i).$$

Specializing to $a_i = i$ shows that $\mathbf{A}_n^{-1}$ is a triple diagonal matrix with elements as indicated:

$$\begin{array}{ll} (i, i)\, i < n & : 4i^3/(4i^2 - 1) \\ (n, n) & : n^2/(2n-1) \\ (i, i+1), (i+1, i) & : -i(i+1)/(2i+1). \end{array}$$

Since $\det \mathbf{B}=1$ we have $\det \mathbf{A}=\det \mathbf{B}\,\mathbf{A}=\prod_{i=1}^{n-1}(r\,b_{n-1}^2)$. In the special case we get

$$\det \mathbf{A}_7=(-1)^{n-1}\cdot\frac{(2n)!}{2^n(n!)^3}\sim\frac{1}{\sqrt{2\pi n}}\left(\frac{2e}{n}\right)^n.$$

(iv) The matrix $\mathbf{A}_8$ is one of the most important in numerical analysis -- it can be called the "second difference matrix" since when we postmultiply $\mathbf{A}_8$ by a vector we get a new vector whose elements are the second differences of the first.

We have already noted (Problem 5.1) that in the case $n=4$ we have

$$\mathbf{A}_8^{-1}=-\frac{1}{5}\begin{bmatrix}4&3&2&1\\3&6&4&2\\2&4&6&2\\1&2&3&4\end{bmatrix}.$$

It can be verified that $\mathbf{A}_8^{-1}$ in the general case has

$$a_{ij}=-i(n-j+1)/(n+1)\qquad i\leqq j,$$

$$a_{ij}=a_{ji}\qquad i>j.$$

We return to a proof of this later. We easily prove by induction that

$$\det \mathbf{A}_8=(-1)^n(n+1).$$

(Compare Problem 4.14.)

We shall now show that the characteristic values of $\mathbf{A}_8$ are, for $k=1,2,\ldots,n$,

$$\alpha_k=-2+2\cos\left(k\pi/(n+1)\right)=-4\sin^2\left(k\pi/(n+1)\right)$$

and the corresponding characteristic vector is

$$c_k=[\sin k\varphi,\ \sin 2k\varphi,\ \sin 3k\varphi,\ \ldots,\ \sin nk\varphi]',$$

where $\varphi=\pi/(n+1)$. This establishes a remark made in (i) above.

It is not much more difficult to handle the case of a general triple diagonal matrix with constant diagonals

$$\mathbf{A}=\begin{bmatrix}a&b&&&\\c&a&b&&\\&&\ddots&&\\&&c&a&b\\&&&c&a\end{bmatrix}.$$

Clearly we have

$$\Delta_n(\lambda)=\det(\mathbf{A}-\lambda\mathbf{I})=(a-\lambda)\Delta_{n-1}(\lambda)-b\,c\,\Delta_{n-2}(\lambda),$$

where

$$\Delta_1=a-\lambda,\quad \Delta_2=(a-\lambda)^2-b\,c.$$

Introduce the notation $D_1=(a-\lambda)/\sqrt{bc}=2\alpha=2\cos\theta$ and $D_2=(a-\lambda)^2/bc-1=4\alpha^2-1=4\cos^2\theta-1$. Generally, write

$$D_n=\Delta_n/\sqrt{(bc)^n}$$

so that

$$D_n=\frac{a-\lambda}{\sqrt{bc}}D_{n-1}-D_{n-2},$$

that is

$$D_n-2\alpha D_{n-1}+D_{n-2}=0,\quad \alpha=\frac{a-\lambda}{2\sqrt{bc}}.$$

We solve this difference equation by trying $D_n=A\mu^n$ and see that μ must satisfy the characteristic equation

$$\mu^2-2\alpha\mu+1=0,$$

so that

$$\mu=\alpha\pm\sqrt{\alpha^2-1}=\begin{cases}c+is\\ c-is,\end{cases}$$

where we suppose $-1\leqq\alpha\leqq1$, and have written $c=\cos\theta=\alpha$, $s=\sin\theta=\sqrt{1-\alpha^2}$. Thus

$$D_n=A_1(c+is)^n+A_2(c-is)^n=B_1\cos n\theta+B_2\sin n\theta.$$

We solve for B_1, B_2 using the initial conditions which give

$$B_1c+B_2s=2c,\quad B_1(2c^2-1)+B_2(2sc)=4c^2-1.$$

Hence $B_1=1$, $B_2=c/s$, so that

$$D_n=\cos n\theta+\frac{\cos\theta}{\sin\theta}\sin n\theta=\frac{\sin(n+1)\theta}{\sin\theta}.$$

Now $\Delta(\lambda)=0$ if and only if $D_n=0$ and so if and only if $\theta=k\pi/(n+1)$, $k=1,2,\ldots,n$. Hence

$$\frac{a-\lambda}{2\sqrt{bc}}=\cos\frac{k\pi}{n+1},\quad \lambda=a-2\sqrt{bc}\cos\frac{k\pi}{n+1}.$$

Since $\cos\frac{k\pi}{n+1}=-\cos\frac{(n+1-k)}{n+1}\pi$ we may take instead

$$\lambda_k=a+2\sqrt{bc}\cos\frac{k\pi}{n+1},\quad k=1,2,\ldots,n.$$

Let us take the case $a=0$, $b=c=1$ for simplicity and determine the characteristic vector corresponding to

$$\lambda_l=2\cos l\theta,\quad \theta=\frac{\pi}{n+1}.$$

The equations read

$$\left.\begin{aligned} x_2&=\lambda_l x_1\\ x_1+x_3&=\lambda_l x_2\\ &\cdots\\ x_{n-1}&=\lambda_l x_n \end{aligned}\right\}.$$

If $x_1=0$ then all the other x's are zeros and this is not allowed in the case of a characteristic vector. We choose $x_1=\sin l\theta$ and find

$$\begin{aligned} x_2&=2\cos l\theta\, x_1=\sin 2l\theta\\ x_3&=\sin 3l\theta\\ &\cdots\\ x_n&=\sin nl\theta. \end{aligned}$$

One way of finding $\mathbf{A}_8^{-1}$ is to use the result of Problem 4.8 in the case where $\mathbf{A}_{22}$ is 1×1 to show how to obtain the inverse of a bordered matrix. We find

$$\begin{bmatrix}\mathbf{A} & \boldsymbol{c}\\ \boldsymbol{r}' & a\end{bmatrix}^{-1}=\begin{bmatrix}\mathbf{A}^{-1}+\alpha\mathbf{A}^{-1}\boldsymbol{c}\boldsymbol{r}'\mathbf{A}^{-1}, & -\alpha\mathbf{A}^{-1}\boldsymbol{c}\\ -\alpha\boldsymbol{r}'\mathbf{A}^{-1}, & \alpha\end{bmatrix}$$

where $\alpha=(a-\boldsymbol{r}'\mathbf{A}^{-1}\boldsymbol{c})^{-1}$, on the assumption that $a\neq\boldsymbol{r}'\mathbf{A}^{-1}\boldsymbol{c}$.

We shall apply this to establish inductively that the inverse of the $n\times n$ matrix:

$$\mathbf{A}_8=\begin{bmatrix}-2 & 1 & & & \\ 1 & -2 & 1 & & \\ & & \cdots & & \\ & & 1 & -2 & 1\\ & & & 1 & -2\end{bmatrix} \quad\text{is}\quad \frac{-1}{n+1}\begin{bmatrix} 1\cdot n & 1\cdot(n-1) & 1\cdot 1\\ 1\cdot(n-1) & 2\cdot(n-1) & 2\cdot 1\\ & & \ddots\\ 1\cdot 1 & 2\cdot 1 & n\cdot 1\end{bmatrix}.$$

Assuming this for a particular n we consider bordering by $\boldsymbol{c}'=\boldsymbol{r}'=[0, 0, \ldots, 0, 1]$, $a=-2$ to get to the next case. Then

$$\alpha^{-1}=-2+\frac{n}{n+1}=\frac{-2n+2+n}{n+1}=\frac{n+2}{n+1}$$

so that $\alpha=\dfrac{-1}{n+2}((n+1)\cdot 1)$ has the appropriate form. Observe that $\boldsymbol{r}'\mathbf{A}^{-1}\boldsymbol{c}$ is just the value of the quadratic form $\boldsymbol{x}'\mathbf{A}^{-1}\boldsymbol{x}$ for $\boldsymbol{x}=[0, 0, \ldots, 0, 1]'$, i.e. $(\mathbf{A}^{-1})_{nn}x_n^2=(\mathbf{A}^{-1})_{nn}$.

In view of symmetry it is enough to consider the row vector $-\alpha\boldsymbol{r}'\mathbf{A}^{-1}$

which is clearly $-\alpha$ times the last row of $\mathbf{A}^{-1}$, i.e.,

$$+\left(\frac{n+1}{n+2}\right)\cdot\frac{-1}{n+1}\cdot[1\cdot 1, 2\cdot 1, \ldots, n\cdot 1]$$

$$=\frac{-1}{n+2}[1\cdot 1, 2\cdot 1, \ldots, n\cdot 1]$$

which is again of the appropriate form.

For reasons of symmetry it will be sufficient to deal with the i, j element when $i \leqq j$.

First note that

$$[\mathbf{A}^{-1}]_{i,j}=\frac{-i(n-j+1)}{n+1}$$

and then compute $[\alpha \mathbf{A}^{-1} \mathbf{c}\, \mathbf{r}' \mathbf{A}^{-1}]_{ij}$ as $\alpha[\mathbf{A}^{-1}\mathbf{c}]_i[\mathbf{r}'\mathbf{A}^{-1}]_j$ giving

$$-\left(\frac{n+1}{n+2}\right)\cdot\left(\frac{-1}{n+1}\cdot(i\cdot 1)\right)\left(\frac{-1}{n+1}\cdot(j\cdot 1)\right)$$

$$=\frac{ij}{(n+1)(n+2)}.$$

Adding we find the new i, j element is

$$\frac{i}{n+1}\left[n-j+1+\frac{j}{n+2}\right]=\frac{-i(n-j+2)}{n+2}$$

which is of the appropriate form.

This completes the proof since the result for $n=1$ is clearly in the appropriate form

$$[-2]^{-1}=-\tfrac{1}{2}[1\cdot 1].$$

(v) This matrix was introduced to us by J. W. Givens. It is easy to verify that $\mathbf{A}_9^{-1}$ is a *triple diagonal matrix* such that

$$\operatorname{diag} \mathbf{A}_9^{-1}=\left[\tfrac{3}{2}, 1, 1, \ldots, 1, 1, \tfrac{1}{2}\right]$$

and the off-diagonal elements are all $-\frac{1}{2}$.

Using results of D. E. Rutherford (Proc. Royal Soc. Edinburgh *62A*, 229—236 (1967); *63A*, 232—241 (1952)) we find that the characteristic roots of $\mathbf{A}_9$ are

$$\frac{1}{2}\sec^2\frac{(2r-1)\pi}{4n}, \quad r=1, 2, \ldots, n.$$

Using the method of p. 30 we find, if $f_n = \det 2\mathbf{A}_9^{-1}$, that

$$f_0 = 1, \quad f_1 = 3, \quad f_r = 2f_{r-1} - f_{r-2}, \ \ldots, \quad r = 2, \ldots, n-1$$

$$f_n = f_{n-1} - f_{n-2}.$$

This gives $f_n = 2$. Hence $\det \mathbf{A}_9^{-1} = 2 \times 2^{-n} = 2^{1-n}$ and

$$\det \mathbf{A}_9 = 2^{n-1}.$$

(vi) M. Fiedler has shown that

$$\mathbf{A}_{14}^{-1} = -\frac{1}{2}\begin{bmatrix} 1-\frac{1}{n-1} & -1 & & & & & \frac{1}{n-1} \\ -1 & 2 & -1 & & & & \\ & -1 & 2 & -1 & & & \\ & & & \ddots & & & \\ & & & -1 & 2 & & -1 \\ -\frac{1}{n-1} & & & & -1 & 1 & -\frac{1}{n-1} \end{bmatrix}$$

and

$$\det \mathbf{A}_{14} = (-1)^{n-1} 2^{n-2} (n-1).$$

Fiedler has handled the case of the matrix

$$\mathbf{C} = [|c_i - c_j|]$$

where c_i's are given constants. He has shown that

$$-2\mathbf{C}^{-1} =$$

$$= \begin{bmatrix} \frac{1}{c_2-c_1} - \frac{1}{c_n-c_1}, & \frac{-1}{c_2-c_1}, & & \frac{-1}{c_n-c_1} \\ -\frac{1}{c_2-c_1}, & \frac{1}{c_2-c_1} + \frac{1}{c_3-c_2}, & \frac{1}{c_3-c_2} & 0 \\ & \ddots & & \\ & \frac{-1}{c_{n-1}-c_{n-2}}, & \frac{1}{c_{n-1}-c_{n-2}} + \frac{1}{c_n-c_{n-1}}, & \frac{-1}{c_n-c_{n-1}} \\ -\frac{1}{c_n-c_1}, & 0 & \frac{-1}{c_n-c_n-1}, & \frac{1}{c_n-c_{n-1}} - \frac{1}{c_n-c_1} \end{bmatrix},$$

and that

$$\det \mathbf{C} = (-1)^{n-1} 2^{n-2} (c_2-c_1)(c_3-c_2) \cdots (c_n-c_1).$$

It is known that $\mathbf{A}_{14}$ has a dominant positive characteristic value and that all the other characteristic values are negative (see e.g. G. Szegö, Amer. Math. Monthly, *43*, 246—259 (1936).)

(vii) This is a particular Vandermonde matrix. The general form is

$$[\mathbf{A}_{15}]_{\lambda\mu} = x_\mu^{\lambda-1}$$

where the x_μ are distinct. The inverse is given by

$$[\mathbf{A}_{15}]^{-1}_{\lambda\mu} = (n+1)^{-1} \exp(-2\pi i(\lambda-1)(\mu-1)/n).$$

(See W. Gautschi, Numer. Math., *4*, 117—123 (1962), and *5*, 425—430 (1963).)

The characteristic values of $\mathbf{A}_{15}$ are

if $n \equiv 0\,(4)$: $n^{1/2}, n^{1/2}, -n^{1/2}, in^{1/2}; \pm n^{1/2}, \pm in^{1/2}$ each $((n/4)-1)$ times

if $n \equiv 1\,(4)$: $n^{1/2}; \pm n^{1/2}, \pm in^{1/2}$ each $(n-1)/4$ times

if $n \equiv 2\,(4)$: $\pm n^{1/2}; \pm n^{1/2}, \pm in^{1/2}$ each $(n-2)/4$ times

if $n \equiv 3\,(4)$: $\pm n^{1/2}, in^{1/2}; \pm n^{1/2}, \pm in^{1/2}$ each $(n-3)/4$ times.

(See L. Carlitz, Acta Arith., *5*, 293—308 (1959).)

It can be shown that

$$\det \mathbf{A}_{15} = n^{(n/2)}.$$

(viii) This is the Hilbert matrix. The case when $n=4$ has been dealt with in Problem 5.5.

There are several ways available to obtain the inverse explicitly. We give an account due to N. Gastinel (Chiffres *3*, 169—152 (1960)). As in most treatments, a more general problem can be handled without complication. Take therefore

$$\mathbf{B} = [b_{ij}] \quad \text{where} \quad b_{ij}^{-1} = b_i - \beta_j \quad i, j = 1, 2, \ldots, n$$

and assume the b's and β's distinct and no b_i coincides with any β_j.

Let

$$b(t) = \prod_i (x-b_i), \quad \beta(t) = \prod_i (x-\beta_i)$$

and write

$$F(t) = b(t) + h(t)$$

where $h(t)$ is a polynomial of degree at most $n-1$. From the elementary theory of partial fractions

$$\frac{F(t)}{\beta(t)} = 1 + \sum_i \frac{A_i}{t-\beta_i} \quad \text{where} \quad A_i = \frac{F(\beta_i)}{\beta'(\beta_i)}. \tag{1}$$

Put $t = b_k$ in (1), to get, since $F(b_k) = h(b_k)$,

$$\frac{h(b_k)}{\beta(b_k)} = 1 + \sum_i \frac{A_i}{b_k - \beta_i}, \quad k = 1, 2, \ldots, n. \tag{2}$$

We can present the equation (2) in matrix-vector form:

$$\mathbf{U}x=y \quad \text{where} \quad x=\begin{bmatrix}-A_1\\-A_2\\ \vdots\\-A_n\end{bmatrix}, \quad y=\begin{bmatrix}1-h(b_1)(\beta(b_1))^{-1}\\1-h(b_2)(\beta(b_2))^{-1}\\ \vdots\\1-h(b_n)(\beta(b_n))^{-1}\end{bmatrix}$$

and $\mathbf{U}$ is a matrix with

$$\mathbf{U}_{k,l}=(b_k-\beta_l)^{-1}.$$

We now observe that in order to solve the system

$$\mathbf{U}x=y$$

for a given y, we should determine a polynomial $h(t)$, of degree at most $n-1$, such that

$$(3) \qquad h(\beta_i)=\beta(b_i)[1-y_i] \quad i=1,2,\ldots,n$$

and then the solution is

$$x_i=-A_i \quad \text{where} \quad A_i=[b(\beta_i)+h(\beta_i)]/\beta'(\beta_i), \quad i=1,2,\ldots,n.$$

In the case considered the data (3) determines the polynomial $h(t)$ by Lagrangian interpolation.

In order to find the inverse $\mathbf{V}$ of $\mathbf{U}$ we have to solve $\mathbf{U}x=y$ in the case when $y_i=\delta_{ij}$, $i=1,2,\ldots,n$, j fixed.

Thus the polynomial $h_j(t)$ is determined by

$$h_j(b_i)=\beta(b_i), \quad i\neq j, \quad h_j(b_j)=0$$

and is given explicitly by

$$h_j(t)=\sum_{k=1}^{n}{}' \frac{\beta(b_k)}{b'(b_k)}\cdot\frac{b(t)}{t-b_k}$$

$$=b(t)\left[\sum\frac{\beta(b_k)}{b'(b_k)}\cdot\frac{1}{t-b_k}-\frac{\beta(b_j)}{b'(b_j)}\cdot\frac{1}{t-b_j}\right].$$

Using in this the fact that

$$\frac{\beta(t)}{b(t)}=1+\sum\frac{\beta(b_k)}{b'(b_k)}\cdot\frac{1}{t-b_k}$$

we find

$$h_j(t)=b(t)\left[\frac{\beta(t)}{b(t)}-1-\frac{\beta(b_j)}{b'(b_j)}\cdot\frac{1}{t-b_j}\right]$$

so that, since $\beta(b_i)=0$,

$$v_{ij}=\frac{-b(\beta_i)+b(\beta_i)\left[-1-\dfrac{\beta(b_j)}{b'(b_j)}\cdot\dfrac{1}{\beta_i-b_j}\right]}{\beta'(\beta_i)}$$

$$=\frac{\beta(b_j)\,b(\beta_i)}{\beta'(\beta_i)\,b'(b_j)(\beta_i-b_j)}.$$

If we put

$$b_i\equiv i,\quad \beta_j=-(j-1)\quad i,j=1,2,\dots,n$$

so that

$$\beta(t)=\prod_0^{n-1}(t+i),\quad b(t)=\prod_1^{n}(t-i)$$

we find

$$b'(b_i)=(-1)^{n-i}(i-1)!\,(n-i)!$$

$$\beta'(\beta_i)=(-1)^{i-1}(i-1)!\,(n-i)!$$

and, finally,

$$[\boldsymbol{A}^{-1}]_{ij}=\frac{(-1)^{i+j}}{i+j-1}\,\frac{(n+i-1)!\,(n+j-1)!}{[(i-1)!\,(j-1)!]^2(n-i)!\,(n-j)!}.$$

For another treatment see S. Schechter, MTAC, *13*, 73—77 (1959). We also find (cf. A. C. Aitken, *Determinants and Matrices* (1951 ed.), p. 136)

$$\det \boldsymbol{A}=\frac{[1!\,2!\,\dots\,(n-1)!]^4}{1!\,2!\,3!\,\dots\,(2n-1)!}.$$

The dominant characteristic values and vectors of $\boldsymbol{H}_n$ have been studied at length (cf. J. Todd, J. Research Nat. Bureau Stand., *65B*, 19—22 (1961)). For the cases when $n=2, 4, 6, 8, 10$ see H. H. Denman and R. C. W. Ettinger, Math. Comp., *16*, 370—371 (1962). We give for $n=4$ the maximum and minimum characteristic values and the corresponding characteristic vectors:

$$1.50021\ 42801,\quad \begin{bmatrix}1. \\ .57017\ 20837 \\ .40677\ 89880 \\ .31814\ 09689\end{bmatrix}$$

$$10^{-4}\times\ .96702\ 30402,\quad \begin{bmatrix}.03688\ 76826 \\ -.41524\ 92878 \\ 1. \\ -.65017\ 12197\end{bmatrix}.$$

(ix) The inverse is the lower triangular matrix given by

$$[\mathbf{A}^{-1}]_{ij}=\begin{cases} 0 & \text{if } j>i \\ \binom{i}{j}\dfrac{B_{i-j}}{i} & \text{if } j\leqq i \end{cases} \quad \text{for} \quad i=1,2,\ldots,$$

where the B's are the Bernoulli numbers defined by

$$B_0=1,\quad B_1=-\frac{1}{2},\quad B_2=\frac{1}{6},\quad B_3=0,\quad B_4=\frac{1}{30},\quad B_5=0,\quad B_6=\frac{1}{42},\ \ldots.$$

Thus in the case $n=6$

$$\mathbf{A}^{-1}=\begin{bmatrix} 1 & 0 & 0 & 0 & 0 & 0 \\ -1/2 & 1/2 & 0 & 0 & 0 & 0 \\ 1/6 & -1/2 & 1/3 & 0 & 0 & 0 \\ 0 & 1/4 & -1/2 & 1/4 & 0 & 0 \\ -1/3 & 0 & 1/3 & -1/2 & 1/5 & 0 \\ 0 & -1/12 & 0 & 5/12 & -1/2 & 1/6 \end{bmatrix}$$

We give two derivations of the expression for the inverse.

(1) Write $\mathbf{A}^{-1}=\mathbf{B}=[\boldsymbol{b}_1, \boldsymbol{b}_2, \ldots, \boldsymbol{b}_n]$ where the $\boldsymbol{b}$'s are column vectors:

$$\boldsymbol{b}_i'=[0, \ldots, 0, b_{ii}, b_{i+1,i}, \ldots, b_{n,i}].$$

Equate the elements in the first column of $\mathbf{AB}=\mathbf{I}$ and we get generally,

$$\sum_{r=1}^{m}\binom{m}{r-1}b_{r1}=0, \quad m=1,2,\ldots,n$$

which are practically the standard recurrence relations for the Bernoulli numbers

$$B_0=1, \quad \sum_{r=0}^{n-1}\binom{n}{r}B_r=0, \quad n=2,3,\ldots. \tag{1}$$

Hence

$$b_{j1}=B_{j-1}, \quad j=1,2,\ldots,n.$$

If we equate elements in the k-th column of $\mathbf{AB}=\mathbf{I}$ we get generally

$$\sum_{j=k}^{l-1}\binom{l}{j-1}b_{jk}=\delta(k,l), \quad l=k,k+1,\ldots,n.$$

If we write $b_{jk}=j^{-1}\binom{j}{k}\beta_{jk}$, $j=k,k+1,\ldots,n$, the above equations give

$$\sum_{l=0}^{m-k}\binom{m-k+1}{l}\beta_{jk}=\delta(k,l)$$

since $m!/(m-k+1)!k!$ cannot vanish. Comparison with (1) gives $\beta_{jk}=B_{j-k}$ and so the result required.

(2) (M. Newman.) Let us denote the k-th column of $\mathbf{B}$ by $[x_0, x_1, \dots x_{n-1}]'$ where $x_0 = x_1 = \dots = x_{k-2} = 0$, $x_{k-1} = k^{-1}$. The (r, k) element of $\mathbf{AB} = \mathbf{I}$ is

$$\sum_{i=0}^{r-1} \binom{r}{i} x_i = \delta(r, k)$$

so that

$$\sum_{i=0}^{r-1} \frac{x_i}{i!\,(r-i)!} = \frac{\delta(r, k)}{r!}. \tag{2}$$

Consider the generating function for the x_r:

$$\omega = \sum_{n=0}^{\infty} \frac{x_n}{n!} z^n. \tag{3}$$

Clearly

$$\omega e^z = \sum_{l=0}^{\infty} \frac{x_l}{l!} z^l \sum_{m=0}^{\infty} \frac{1}{m!} z^m = \sum_{n=0}^{\infty} \left\{ \sum_{l=0}^{n} \frac{x_l}{l!} \frac{1}{(n-l)!} \right\} z^n$$

$$= \sum_{n=0}^{\infty} \left\{ \frac{x_n}{n!} + \frac{\delta(n, k)}{n!} \right\} z^n$$

where we have used (2) at the last step. We have thus obtained

$$\omega e^z = \omega + \frac{z^k}{k!}$$

giving

$$\omega = \frac{1}{k!} \frac{z^k}{e^z - 1} = \frac{z^{k-1}}{k!} \sum_{n=0}^{\infty} \frac{B_n}{n!} z^n \tag{4}$$

in virtue of a standard generating relation for the Bernoulli numbers

$$\frac{z}{e^z - 1} = \sum_{n=0}^{\infty} \frac{B_n}{n!} z^n.$$

Comparison of (3) and (4) gives the result required.

5.15. Solution

We begin by examining the inequality

$$\frac{\|\boldsymbol{\xi}\|}{\|\boldsymbol{x}\|} \leqq \varkappa(\mathbf{W}) \frac{\|\boldsymbol{\beta}\|}{\|\boldsymbol{b}\|} \tag{1}$$

in the case of the Chebyshev norm and the data of Problem 5.3 (and their solutions). We find

$$\frac{\|\boldsymbol{\xi}\|_\infty}{\|\boldsymbol{x}\|_\infty} = \frac{1.36}{1}, \quad \frac{\|\boldsymbol{\beta}\|_\infty}{\|\boldsymbol{b}\|_\infty} = \frac{0.01}{33}, \quad \varkappa_\infty(\mathbf{W}) = 33 \times 136 = 4488$$

and so the inequality (1) is a best possible one in the Chebyshev case.

If we use instead the Manhattan norm we find

$$\frac{\|\boldsymbol{\xi}\|_1}{\|\boldsymbol{x}\|_1}=\frac{2.74}{4}, \quad \frac{\|\boldsymbol{\beta}\|_1}{\|\boldsymbol{b}\|_1}=\frac{.04}{119}, \quad \varkappa_1(\mathbf{A})=\varkappa_\infty(\mathbf{A})$$

so that there is an overestimate in (1) by a factor of about 2 since the error amplification is actually $(2.74\times 119)/(4\times .06)\doteqdot 2038$.

Referring to the proof of (1) it is clear that in order to obtain equality in (1) we must have equality in *both* the relations

$$\|\boldsymbol{b}\|=\|\mathbf{A}\|\,\|\boldsymbol{x}\|, \quad \|\boldsymbol{\xi}\|=\|\mathbf{A}^{-1}\|\,\|\boldsymbol{\beta}\|.$$

Since we have an example where equality holds in (1) in the Chebyshev case we proceed to the Manhattan case. From Chapter 3 we have

$$\|\boldsymbol{b}\|_1=\|\mathbf{A}\|_1\,\|\boldsymbol{x}\|_1$$

if $\boldsymbol{x}=[0, 0, \ldots, 1, \ldots 0]'$ where the 1 is in the j-th place, where the j-th column sum is maximum.

We have in the case of $\mathbf{W}$, $j=3$ and so we have

$$\mathbf{W}\begin{bmatrix}0\\0\\1\\0\end{bmatrix}=\begin{bmatrix}8\\6\\10\\9\end{bmatrix}.$$

In the case of $\mathbf{W}^{-1}$, $j=2$ and so we have

$$\mathbf{W}^{-1}\begin{bmatrix}0\\1\\0\\0\end{bmatrix}=\begin{bmatrix}-41\\68\\-17\\10\end{bmatrix}.$$

Thus if we consider the system

$$\mathbf{W}\boldsymbol{x}=\begin{bmatrix}8\\6\\10\\9\end{bmatrix} \quad \text{with solution} \quad \begin{bmatrix}0\\0\\1\\0\end{bmatrix}$$

and perturb it to

$$\mathbf{W}\boldsymbol{x}=\begin{bmatrix}8\\6+\varepsilon\\10\\9\end{bmatrix} \quad \text{with solution} \quad \begin{bmatrix}0\\0\\1\\0\end{bmatrix}+\varepsilon\begin{bmatrix}-41\\68\\-17\\10\end{bmatrix}$$

we have

$$\|\boldsymbol{\xi}\|_1=136\,\varepsilon, \quad \|\boldsymbol{x}\|_1=1, \quad \|\boldsymbol{\beta}\|_1=\varepsilon, \quad \|\boldsymbol{b}\|_1=33$$

and we have equality in the relation

$$\|\xi\|_1/\|x\|_1 \leqq \varkappa_1(\mathbf{W})\,\|\beta\|_1/\|b\|_1 .$$

The corresponding problem for the case of the euclidean vector norm requires some calculations. Let us first observe what happens in the case of the first perturbation [.01, − .01, .01, − .01]′. We find that

$$\|\beta\| = .02, \quad \|b\| = (32^2+23^3+33^2+31^2)^{1/2} \doteqdot 60,$$

$$\|\xi\| = (82^2+136^2+35^2+21^2)^{1/2} \times .01 \doteqdot 1.64, \quad \|x\| = 2,$$

and, since the characteristic values of $\mathbf{W}$ are approximately

$$30.29, \quad 3.858, \quad .8431, \quad .01015,$$

we have

$$\varkappa_2(\mathbf{W}) \doteqdot 30.29/\ .01015 \doteqdot 2974.$$

Thus there is a rather small overestimate in (1) since the error amplification is actually $(1.64\times 60)/(2\times .02) = 2460$.

Let us now find a b and a β for which there is equality in the euclidean case. We recall that equality in

$$\|\mathbf{A}x\|_2 = \|\mathbf{A}\|_2\,\|x\|_2$$

is attained if x is the dominant characteristic vector of $\mathbf{A}\mathbf{A}'$. In the symmetric case $\|\mathbf{A}\|_2$ is just the (absolute value of the) dominant characteristic root of $\mathbf{A}$ and equality is obtained for the dominant characteristic vector of $\mathbf{A}$.

Suppose $\mathbf{W}$ has dominant characteristic root $\lambda_{\#}$ with vector $v_{\#}$ and that its (absolutely) least characteristic root is λ_b with vector v_b and that $v_{\#}$, v_b have euclidean norm 1.

Consider $\mathbf{W}x = v_{\#}$ perturbed to

$$\mathbf{W}(x+\xi) = v_{\#} + \varepsilon v_b .$$

Then

$$x = v_{\#}/\lambda_{\#}$$

and

$$\xi = \varepsilon v_b/\lambda_b$$

so that

$$\frac{\|\xi\|}{\|x\|} = \frac{\lambda_{\#}}{\varepsilon\lambda_b}, \quad \frac{\|\beta\|}{\|b\|} = \varepsilon.$$

This means that the amplification of the relative error is $\lambda_{\#}/\lambda_b = \varkappa_2(\mathbf{W})$.

We shall only examine $\mathbf{C}_4$ in the euclidean vector norm case. We have seen (Problem 5.1)

$$\varkappa_2(\mathbf{C}_4) = \cot^2(\pi/10) \sim 9.47$$

so that the largest amplification of relative error in this case is about 1/300 of that in the $\mathbf{W}$ case.

Repetition of the argument in $\mathbf{W}$ case shows that this worst case is attained when we consider the system

$$\mathbf{C}_4 \mathbf{x} = \mathbf{v}_{\#}$$

perturbed to

$$\mathbf{C}_4 \mathbf{x} = \mathbf{v}_{\#} + \varepsilon \mathbf{v}_b$$

where $\mathbf{v}_{\#}$, $\mathbf{v}_b$ are the characteristic vectors of $\mathbf{C}_4$ corresponding to the characteristic values

$$-4\cos^2\frac{\pi}{10} = -2(c+1), \quad -4\sin^2\frac{\pi}{10} = -2(c-1)$$

and these can be taken as

$$[1, \ -2c, \ 2c, \ -1]'$$

and

$$[1, \ 2c, \ 2c, \ 1]'$$

where $c = \cos \pi/5$.

We give the result for the $\mathbf{H}_4$ case with the Manhattan norm. The system $\mathbf{H}\mathbf{x} = [1, \frac{1}{2}, \frac{1}{3}, \frac{1}{4}]'$ with solution $\mathbf{x} = [1, 0, 0, 0]'$ when perturbed to $\mathbf{H}\mathbf{x} = [1, \frac{1}{2}, \frac{1}{3} + \varepsilon, \frac{1}{4}]'$ gives as solution

$$\mathbf{x} = [1, \ 0, \ 0, \ 0]' + \varepsilon[240, \ -2700, \ 6480, \ -4200]'$$

and the extreme amplification $\varkappa_1(\mathbf{H}) = (25/12) \times 13620 = 28475$ is attained.

5.16. Solution

Multiply the right-hand side by $\mathbf{A} + \mathbf{x}\mathbf{y}'$ to get

$$\mathbf{I} - \frac{\mathbf{x}\mathbf{y}'\mathbf{A}^{-1}}{1 + \mathbf{y}'\mathbf{A}^{-1}\mathbf{x}} + \mathbf{x}\mathbf{y}'\mathbf{A}^{-1} - \frac{\mathbf{x}\mathbf{y}'\mathbf{A}^{-1}\mathbf{x}\mathbf{y}'\mathbf{A}^{-1}}{1 + \mathbf{y}'\mathbf{A}^{-1}\mathbf{x}}$$

$$= \mathbf{I} + \frac{1}{1 + \mathbf{y}'\mathbf{A}^{-1}\mathbf{x}} \left\{ \begin{array}{c} -\mathbf{x}\mathbf{y}'\mathbf{A}^{-1} \\ + \mathbf{x}\mathbf{y}'\mathbf{A}^{-1} + (\mathbf{x}\mathbf{y}'\mathbf{A}^{-1})(\mathbf{y}'\mathbf{A}^{-1}\mathbf{x}) \\ -\mathbf{x}\mathbf{y}'\mathbf{A}^{-1}\mathbf{x}\mathbf{y}'\mathbf{A}^{-1} \end{array} \right\}.$$

The first two terms in the braces cancel and so do the second two for $\mathbf{y}'\mathbf{A}^{-1}\mathbf{x}$ is a scalar and these two terms are

$$(\mathbf{y}'\mathbf{A}^{-1}\mathbf{x})(\mathbf{x}\mathbf{y}'\mathbf{A}^{-1}) \quad \text{and} \quad -\mathbf{x}(\mathbf{y}'\mathbf{A}^{-1}\mathbf{x})\mathbf{y}'\mathbf{A}^{-1} = -(\mathbf{y}'\mathbf{A}^{-1}\mathbf{x})(\mathbf{x}\mathbf{y}'\mathbf{A}^{-1}).$$

This establishes the relation.

The change in $\mathbf{A}^{-1}$ caused by the change $\mathbf{x}\mathbf{y}'$ in $\mathbf{A}$ is

$$(1 + \mathbf{y}'\mathbf{A}^{-1}\mathbf{x})^{-1}[\mathbf{A}^{-1}\mathbf{x}\mathbf{y}'\mathbf{A}^{-1}].$$

The first factor here is a scalar and can be computed in $2n^2$ multiplications, $\mathbf{A}^{-1}$ being available, by first taking $\mathbf{A}^{-1}\mathbf{x}$ and then $\mathbf{y}'(\mathbf{A}^{-1}\mathbf{x})$. Now using the column vector $\mathbf{A}^{-1}x$ just computed and computing the row vector $x'\mathbf{A}^{-1}$ at the expense of another n^2 multiplications, we obtain the antiscalar product $(\mathbf{A}^{-1}x)(y'\mathbf{A}^{-1})$ at the cost of n^2 more multiplications. Thus in all we need about $4n^2$ multiplications. The scalar factor can be incorporated in the first factor $\mathbf{A}^{-1}x$ at the cost of n divisions.

$$\mathbf{A}+xy'=\begin{bmatrix}3&0&1&1\\0&3&0&1\\1&0&3&0\\1&1&0&3\end{bmatrix},\quad \mathbf{A}^{-1}=\frac{1}{5}\begin{bmatrix}4&3&2&1\\3&6&4&2\\2&4&6&3\\1&2&3&4\end{bmatrix},$$

$$y'\mathbf{A}^{-1}x=10,\quad \mathbf{A}^{-1}x=\begin{bmatrix}2\\3\\3\\2\end{bmatrix},\quad (1+y'\mathbf{A}^{-1}x)^{-1}(\mathbf{A}^{-1}x)=\begin{bmatrix}2/11\\3/11\\3/11\\2/11\end{bmatrix},$$

$$y'\mathbf{A}^{-1}=[2\ \ 3\ \ 3\ \ 2].$$

The perturbation is

$$\frac{1}{11}\begin{bmatrix}4&6&6&4\\6&9&9&6\\6&9&9&6\\4&6&6&4\end{bmatrix},$$

giving

$$\begin{bmatrix}3&0&1&1\\0&3&0&1\\1&0&3&0\\1&1&0&3\end{bmatrix}^{-1}=\frac{1}{55}\begin{bmatrix}24&3&-8&-9\\3&21&-1&-8\\-8&-1&21&3\\-9&-8&3&24\end{bmatrix}.$$

5.17. Solution

(a)

$$\left.\begin{aligned}10^{-2}x+y&=1\\x+y&=2\end{aligned}\right\}$$

Subtracting $$\frac{99}{100}x=1,\quad x=1+\frac{1}{99}$$

$$y=1-\frac{1}{99}.$$

(b)

(1) $$.10\times10^{-1}x+\ .10\times10^{1}y=\ .10\times10^{1}$$

(2) $$.10\times10\ {}^{1}x+\ .10\times10^{1}y=\ .20\times10^{1}.$$

Multiply (1) by 10^2 to get

(3) $$.10\times10^{1}x+\ .10\times10^{3}y=\ .10\times10^{3}.$$

Subtract (3) from (2) to get

$$.10\times 10^3 y = .10\times 10^3$$

so that $y = .10\times 10^1$ and if we substitute this in (1) we get $x=0$.

(c) Multiply (2) by 10^{-2} to get

$$(4) \qquad .10\times 10^{-1}x + .10\times 10^{-1}y = .20\times 10^{-1}.$$

Subtract (4) from (1) to get

$$y = .10\times 10^1$$

and if we substitute this in (2) we get

$$x = .10\times 10^1.$$

5.18. Solution

Our first pivot is the (1, 1) element and we get

$$\left.\begin{array}{r} 10x_1 + 7x_2 + 8x_3 + 7x_4 = 32 \\ .1x_2 + .4x_3 + .1x_4 = .6 \\ .4x_2 + 3.6x_3 + 3.4x_4 = 7.4 \\ .1x_2 + 3.4x_3 + 5.1x_4 = 8.6 \end{array}\right\}.$$

The next pivot is the (4, 4) element and the next stage of reduction is to

$$\left.\begin{array}{l} .10x_2 + .33x_3 \qquad\qquad = .42 \\ .33x_2 + 1.32x_3 \qquad\quad = 1.64 \\ .1\ x_2 + 3.4\ x_3 + 5.1x_4 = 8.6 \end{array}\right\}.$$

The next pivot is the (3, 3) element and the next stage of reduction is to

$$\left.\begin{array}{l} .02x_2 \qquad\qquad = .01 \\ .33x_2 + 1.32x_3 = 1.64 \end{array}\right\}.$$

Thus we get

$$x_2 = .50, \quad x_3 = 1.12, \quad x_4 = -.93, \quad x_1 = 1.30.$$

5.19. Solution

Choose $\mathbf{F}$ so that $\mathbf{A}+\mathbf{F}$ is singular. Then choose any $\mathbf{x}_0 \neq 0$ such that $(\mathbf{A}+\mathbf{F})\mathbf{x}_0 = 0$. Then

$$\|\mathbf{F}\| \geq \frac{\|\mathbf{F}\mathbf{x}_0\|}{\|\mathbf{x}_0\|} = \frac{\|\mathbf{A}\mathbf{x}_0\|}{\|\mathbf{x}_0\|} = \frac{\|\mathbf{A}\mathbf{x}_0\|}{\|\mathbf{A}^{-1}(\mathbf{A}\mathbf{x}_0)\|} \geq \frac{1}{\|\mathbf{A}^{-1}\|}.$$

We show that there is an $\mathbf{F}_0$ such that $\mathbf{A}+\mathbf{F}_0$ is singular and $\|\mathbf{F}_0\| = 1/\|\mathbf{A}^{-1}\|$. We begin by choosing a $\mathbf{y} \neq 0$ such that

$$(1) \qquad \|\mathbf{A}^{-1}\mathbf{y}\|_p = \|\mathbf{A}^{-1}\|_p \|\mathbf{y}\|_p.$$

We then choose a vector ω such that there is equality in the Hölder inequality:

$$|\omega'(A^{-1}y)| \leq \|\omega\|_q \|A^{-1}y\|_p \tag{2}$$

and normalize it so that

$$\omega' A^{-1} y = 1. \tag{3}$$

We then take $F_0 = -y\omega'$. We have (by (3))

$$(A+F_0)A^{-1}y = y - y\omega' A^{-1} y = 0$$

so that $A+F_0$ is singular. We now have

$\|F_0\|_p = \sup_{x \neq 0} \dfrac{\|y\omega' x\|_p}{\|x\|_p}$ by definition of F_0, definition of an induced norm and by homogeneity of a vector norm

$= \sup_{x \neq 0} \dfrac{\|y\|_p |\omega' x|}{\|x\|_p}$ by homogeneity of a vector norm

$= \|y\|_p \sup_{x \neq 0} \dfrac{|\omega' x|}{\|x\|_p}$ since $\|y\|_p$ does not depend on x

$= \|y\|_p \|\omega\|_q$ by Hölder's inequality

$= \|y\|_p / \|A^{-1}y\|_p$ by (2)

$= 1/\|A^{-1}\|_p$. by (1).

The results just established can be stated as follows:

"$\min \|F\|_p = \dfrac{1}{\|A^{-1}\|_p}$ where the minimum is over all F for which $A+F$ is singular".

We can interpret this result as follows: "If a matrix A has a large condition number it is 'near' to a singular matrix". This result is true for a general norm and the proof is unchanged except that the q-norms must be replaced by the norm dual to that given.

We carry out the detailed calculations of this problem in the case of the matrix A of Problem 5.6 in the case of the Chebyshev norm, i.e. $p = \infty$.

With $A = \begin{bmatrix} 1 & 1 \\ 1 & 1-n^{-2} \end{bmatrix}$ we have $A^{-1} = \begin{bmatrix} 1-n^2 & n^2 \\ n^2 & -n^2 \end{bmatrix}$ and

$$\|A\|_\infty = 2, \quad \|A^{-1}\|_\infty = 2n^2.$$

We therefore ought to be able to find an F with $\|F\|_\infty = (2n^2)^{-1}$ such that $A+F$ is singular. If we take

$$F = \begin{bmatrix} 0 & 0 \\ 0 & n^{-2} \end{bmatrix}$$

$\boldsymbol{A}+\boldsymbol{F}$ is clearly singular but $\|\boldsymbol{F}\|=n^{-2}$. However, if we take

$$\boldsymbol{F}_0=\begin{bmatrix}0 & -(2n^2)^{-1}\\ 0 & (2n^2)^{-1}\end{bmatrix}$$

then $\boldsymbol{A}+\boldsymbol{F}_0$ is singular and $\|\boldsymbol{F}_0\|_\infty=(2n^2)^{-1}$.

To get $\boldsymbol{F}_0$ by the method sketched above as we begin by choosing $\boldsymbol{y}=[1,-1]'$ which gives

$$\boldsymbol{A}^{-1}\boldsymbol{y}=[1-2n^2, 2n^2]', \quad \|\boldsymbol{A}^{-1}\boldsymbol{y}\|_\infty=2n^2.$$

We write down the appropriate Hölder inequality

$$|\boldsymbol{w}'(\boldsymbol{A}^{-1}\boldsymbol{y})|\leqq\|\boldsymbol{w}\|_1\|\boldsymbol{A}^{-1}\boldsymbol{y}\|_\infty$$

in which there is equality if

$$|w_1(1-2n^2)+w_2(2n^2)|=(|w_1|+|w_2|)2n^2$$

and for this we must have $w_1=0$, $w_2=1$.

In order to satisfy the normalizing condition (3) we must take $w_1=0$, $w_2=(2n^2)^{-1}$.

Hence

$$\boldsymbol{F}_0=-\begin{bmatrix}1\\-1\end{bmatrix}[0,(2n^2)^{-1}]=\begin{bmatrix}0, & -(2n^2)^{-1}\\ 0, & (2n^2)^{-1}\end{bmatrix}.$$

5.20. Solution

(a) From $\boldsymbol{Ax}=\boldsymbol{b}$ and $(\boldsymbol{A}+\boldsymbol{F})(\boldsymbol{x}+\xi)=\boldsymbol{b}$ it follows that

$$(\boldsymbol{A}+\boldsymbol{F})\xi=-\boldsymbol{Fx} \quad \text{or} \quad \xi=-(1+\boldsymbol{A}^{-1}\boldsymbol{F})^{-1}\boldsymbol{A}^{-1}\boldsymbol{Fx}$$

and this gives

$$\|\xi\|\leqq\|(\boldsymbol{I}+\boldsymbol{A}^{-1}\boldsymbol{F})^{-1}\|\,\|\boldsymbol{A}^{-1}\boldsymbol{F}\|\,\|\boldsymbol{x}\|.$$

If we use the norm axioms we find

$$\|\boldsymbol{A}^{-1}\boldsymbol{F}\|\leqq\|\boldsymbol{A}^{-1}\|\,\|\boldsymbol{F}\|,$$

$$\|(\boldsymbol{I}+\boldsymbol{A}^{-1}\boldsymbol{F})^{-1}\|\leqq(\|\boldsymbol{I}+\boldsymbol{A}^{-1}\boldsymbol{F}\|)^{-1},$$

$$\|\boldsymbol{I}\|=\|\boldsymbol{I}+\boldsymbol{A}^{-1}\boldsymbol{F}-\boldsymbol{A}^{-1}\boldsymbol{F}\|\leqq\|\boldsymbol{I}+\boldsymbol{A}^{-1}\boldsymbol{F}\|+\|\boldsymbol{A}^{-1}\boldsymbol{F}\|,$$

so that

$$\|\boldsymbol{I}+\boldsymbol{A}^{-1}\boldsymbol{F}\|\geqq1-\|\boldsymbol{A}^{-1}\boldsymbol{F}\|\geqq1-\|\boldsymbol{A}^{-1}\|\,\|\boldsymbol{F}\|.$$

Hence

$$\|\xi\|\leqq\frac{\|\boldsymbol{A}^{-1}\|\,\|\boldsymbol{F}\|\,\|\boldsymbol{x}\|}{1-\|\boldsymbol{A}^{-1}\|\,\|\boldsymbol{F}\|}$$

and this gives the result stated.

(b) From the identity

$$\boldsymbol{A}(\boldsymbol{A}^{-1}-\boldsymbol{B}^{-1})\boldsymbol{B}=\boldsymbol{B}-\boldsymbol{A}$$

we get

$$(\mathbf{A}^{-1}-\mathbf{B}^{-1})=\mathbf{A}^{-1}(\mathbf{B}-\mathbf{A})\mathbf{B}^{-1}$$

which gives

$$\|\mathbf{A}^{-1}-\mathbf{B}^{-1}\| \leqq \|\mathbf{A}^{-1}\|\,\|\mathbf{B}-\mathbf{A}\|\,\|\mathbf{B}^{-1}\|$$

so that, as required,

$$\frac{\|\mathbf{A}^{-1}-\mathbf{B}^{-1}\|}{\|\mathbf{B}\|} \leqq \varkappa(\mathbf{A})\frac{\|\mathbf{B}-\mathbf{A}\|}{\|\mathbf{A}\|}.$$

5.21. Solution

The matrix $\mathbf{A}_n$ was suggested by H. Rutishauser as a test matrix for matrix inversion programs in view of its ill-condition.

The i-th row of $\mathbf{B}_n$ is

$$\binom{i-1}{0},\ -\binom{i-1}{1},\ \binom{i-1}{2},\ \dots,\ (-1)^{i-1}\binom{i-1}{i-1},\ 0,\ \dots,\ 0,$$

and the j-th column of $\mathbf{B}_n'$ is

$$\binom{j-1}{0},\ -\binom{j-1}{1},\ \binom{j-1}{2},\ \dots,\ (-1)^{j-1}\binom{j-1}{j-1},\ 0,\ \dots,\ 0,$$

so that the ij element of $\mathbf{A}_n$ is

$$\binom{i-1}{0}\binom{j-1}{0}+\binom{i-1}{1}\binom{j-1}{1}+\dots$$

$$=\binom{i-1}{0}\binom{j-1}{j-1}+\binom{i-1}{1}\binom{j-1}{j-2}+\dots$$

which is the coefficient of x^{j-1} in the product $(1+x)^{i-1}(1+x)^{j-1}$ which is $\binom{i+j-2}{j-1}$.

This method can be applied to prove $\mathbf{B}_n^2=\mathbf{I}$. Thus

$$\mathbf{A}_4=\begin{bmatrix}1&1&1&1\\1&2&3&4\\1&3&6&10\\1&4&60&20\end{bmatrix}.$$

Observe that since $\mathbf{B}^2=\mathbf{I}$ we have $\mathbf{A}^{-1}=\mathbf{B}'\mathbf{B}$ and so $\mathbf{A}=\mathbf{B}\mathbf{A}^{-1}\mathbf{B}^{-1}$, i.e. $\mathbf{A}$, $\mathbf{A}^{-1}$ are similar, so that their characteristic roots are the same.

5.22. Solution

(See G. J. Tee, SIGNUM Newsletter, *7*, 19—20, (1972), and G. Zielke, ibid., *9*, 11—12 (1974).)

Note that

$$\mathbf{A}^{-1}=\begin{bmatrix}-557 & 842 & -284\\ 610 & -922 & 311\\ 2 & -3 & 1\end{bmatrix}$$

so that $\|\mathbf{A}\|_\infty=93$, $\|\mathbf{A}^{-1}\|_\infty=1843$. Tee does not discuss the worst case in the context of the effect of the change in $\boldsymbol{x}$ caused by a change in $\boldsymbol{b}$. We discuss this in the case of the Chebyshev norm following Problem 5.15. If we take

$$\boldsymbol{b}=\begin{bmatrix}7\\36\\93\end{bmatrix},\quad \boldsymbol{x}=\begin{bmatrix}1\\1\\-1\end{bmatrix},\quad \delta\boldsymbol{b}=\begin{bmatrix}1\\-1\\1\end{bmatrix}$$

then

$$\delta\boldsymbol{x}=\begin{bmatrix}-1643\\1843\\6\end{bmatrix}$$

and

$$[\|\delta\boldsymbol{x}\|_\infty/\|\boldsymbol{x}\|_\infty]/[\|\delta\boldsymbol{b}\|_\infty/\|\boldsymbol{b}\|_\infty]=1843\times 93=171\,399.$$

The Zielke matrix $\mathbf{Z}$ is got by a rank-one perturbation of $\mathbf{A}$ since $\mathbf{Z}=\mathbf{A}+\alpha\mathbf{J}$ where $\mathbf{J}=\boldsymbol{e}\boldsymbol{e}'$ with $\boldsymbol{e}'=[1, 1, 1]$. We can therefore use the result of Problem 5.16 to find

$$\mathbf{Z}^{-1}=\mathbf{A}^{-1}-\frac{\alpha\mathbf{A}^{-1}\boldsymbol{e}\boldsymbol{e}'\mathbf{A}^{-1}}{1+\alpha\boldsymbol{e}'\mathbf{A}^{-1}\boldsymbol{e}}.$$

Here $\boldsymbol{e}'\mathbf{A}^{-1}=[55, -83, 28]$, $\mathbf{A}^{-1}\boldsymbol{e}=[1, -1, 0]'$ and $\boldsymbol{e}'\mathbf{A}^{-1}\boldsymbol{e}=0$. Hence

$$\begin{aligned}\mathbf{Z}^{-1}&=\mathbf{A}^{-1}-\alpha[1,\ -1,\ 0]'[55,\ -83,\ 28]\\ &=\begin{bmatrix}-55\alpha-557 & 83\alpha+842 & -28\alpha-284\\ 55\alpha+610 & -83\alpha+922 & 28\alpha+311\\ 2 & -3 & 2\end{bmatrix}.\end{aligned}$$

When α is large and positive we have

$$\|\mathbf{Z}\|_\infty=35+3\alpha,\quad \|\mathbf{Z}^{-1}\|_\infty=1843+166\alpha.$$

If we take

$$\boldsymbol{b}=\begin{bmatrix}3\alpha+35\\3\alpha+10\\3\alpha-39\end{bmatrix},\quad \boldsymbol{x}=\begin{bmatrix}1\\1\\1\end{bmatrix},\quad \delta\boldsymbol{b}=\begin{bmatrix}1\\-1\\1\end{bmatrix}$$

we find

$$\delta\boldsymbol{x}=\begin{bmatrix}-166\alpha-1683\\166\alpha+1843\\6\end{bmatrix}.$$

Now we have

$$[\|\delta x\|_\infty/\|x\|_\infty]/[\|\delta b\|_\infty/\|b\|_\infty]=(166\alpha+1843)(3\alpha+35)\sim 498\alpha^2.$$

5.23. Solution

See O. Taussky, MTAC, *4*, 111—112 (1950).

Generally the condition of $\boldsymbol{A}\boldsymbol{A}'$ is worse than that of $\boldsymbol{A}$ so that symmetrizing a system of equations, apart from its expense, makes the solution more awkward.

Chapter 6

6.1. Solution

If $r'A=\alpha r'$ and $Ac=\beta c$ then $r'Ac=(r'A)c=\alpha r'c$ and $r'Ac=r'(Ac)=\beta r'c$. Hence

$$(\alpha-\beta)r'c=0$$

and so

$$r'c=0$$

since $\alpha\neq\beta$.

6.2. Solution

xy' has rank 1 if $x\neq 0$, $y\neq 0$. Its characteristic values are $x'y$ and 0 with multiplicity $n-1$. We illustrate this in a 3×3 case. The characteristic polynomial in the case $x=[1,k_1,k_2]'$, $y=[a,b,c]'$ is

$$\det\begin{bmatrix} a-\lambda & b & c \\ k_1 a & k_1 b-\lambda & k_1 c \\ k_2 a & k_2 b & k_2 c-\lambda \end{bmatrix}$$

$$=\det\begin{bmatrix} a+k_1b+k_2c-\lambda & b & c \\ k_1a+k_1^2b+k_1^2c-k_1\lambda & k_1b-\lambda & k_1c \\ k_2a+k_2^2b+k_2^2c-k_2\lambda & k_2b & k_2c-\lambda \end{bmatrix} \text{ by adding } k_1\,\text{col}_2+k_2\,\text{col}_3 \text{ to col}_1$$

$$=(a+k_1b+k_2c-\lambda)\det\begin{bmatrix} 1 & b & c \\ k_1 & k_1b-\lambda & k_1c \\ k_2 & k_2b & k_2c-\lambda \end{bmatrix}$$

$$=(a+k_1b+k_2c-\lambda)\det\begin{bmatrix} 1 & b & c \\ 0 & -\lambda & 0 \\ 0 & 0 & -\lambda \end{bmatrix} \text{ by row}_2-k_1\,\text{row}_1 \text{ and row}_3-k_2\,\text{row}_1.$$

We can apply this result with

$$x'=[k_1,k_2,\ldots,k_n],\quad y'=[k_1^{-1},k_2^{-1},\ldots,k_n^{-1}]$$

and

$$x'y=1+1+\ldots+1=n.$$

Alternatively, the matrix can be represented a

$$\mathbf{D}^{-1}\mathbf{J}\mathbf{D}$$

when $\mathbf{D}=\text{diag}\,[k_1, k_2, \ldots, k_n]$ and we can use the result of Problem 5.13 (ii).

6.3. Solution

$\mathbf{A}=\begin{bmatrix}0 & 1\\ 0 & 0\end{bmatrix}$ and $\mathbf{B}=\begin{bmatrix}0 & 0\\ 1 & 0\end{bmatrix}$ each have characteristic values 0, 0 but

$\mathbf{A}+\mathbf{B}=\begin{bmatrix}0 & 1\\ 1 & 0\end{bmatrix}$ has characteristic values ± 1, and

$\mathbf{AB}=\begin{bmatrix}1 & 0\\ 0 & 0\end{bmatrix}$ has characteristic values 1, 0.

A simple condition is that $\mathbf{A}$ is a polynomial in $\mathbf{B}$ (see Problem 6.4 below).

6.4. Solution

Clearly if $\mathbf{A}\boldsymbol{a}=\alpha\boldsymbol{a}$ then $\mathbf{A}^2\boldsymbol{a}=\mathbf{A}(\mathbf{A}\boldsymbol{a})=\mathbf{A}(\alpha\boldsymbol{a})=\alpha(\mathbf{A}\boldsymbol{a})=\alpha^2\boldsymbol{a}$ and, generally,

$$p(\mathbf{A})\,\boldsymbol{a}=p(\alpha)\,\boldsymbol{a}.$$

Also $\mathbf{A}\boldsymbol{a}=\alpha\boldsymbol{a}$ implies $\alpha^{-1}\boldsymbol{a}=\mathbf{A}^{-1}\boldsymbol{a}$ if $\mathbf{A}$ is non-singular.

The same result is true

$$r(\mathbf{A})\,\boldsymbol{a}=r(\alpha)\,\boldsymbol{a}$$

with the proviso that $r(\mathbf{A})$ is not singular.

6.5. Solution

It is clear that the characteristic polynomial of $\mathbf{P}$ is x^4-1. Accordingly its characteristic values are $\zeta_4^r=\exp(\pi i r/2)$ for $r=1, 2, 3, 4$. Corresponding to the characteristic value ζ_4^r we have a characteristic vector

$$\boldsymbol{v}_r=[\zeta_4^r, \zeta_4^{2r}, \zeta_4^{3r}, \zeta_4^{4r}]'.$$

The characteristic values of $\mathbf{Q}$ are (compare Problem 6.4)

$$a+b\zeta_4^r+c\zeta_4^{2r}+d\zeta_4^{3r}, \quad r=1, 2, 3, 4$$

and the characteristic vectors *include* the $\boldsymbol{v}_r$. (Note that there can be more, e.g. if $a=1$, $b=c=d=0$, any vector is a characteristic vector of $\mathbf{Q}=\mathbf{I}$.)

The extension to the general case is obvious.

Applying these results to the matrix $\mathbf{R}$ we see that its characteristic values are, for $r=1, 2, 3, 4$,

$$\begin{aligned}
&-2+1\cdot\exp(r\pi i/2)+0\cdot\exp(r\pi i)+1\cdot\exp(3\pi i r/2)\\
&=-2+[\cos(r\pi/2)+i\sin(r\pi/2)]+[\cos(3r\pi/2)+i\sin(3r\pi/2)]\\
&=-2+2\cos(r\pi/2)\\
&=-4\sin^2(r\pi/4)
\end{aligned}$$

i.e., $\quad -2,\ -4,\ -2,\ 0.$

6.6. Solution

Suppose

$$\begin{bmatrix} \mathbf{A} & \mathbf{B} \\ \mathbf{B} & \mathbf{A} \end{bmatrix}\begin{bmatrix} x_1 \\ x_2 \end{bmatrix} = \lambda \begin{bmatrix} x_1 \\ x_2 \end{bmatrix}.$$

Then

$$\mathbf{A}\mathbf{x}_1 + \mathbf{B}\mathbf{x}_2 = \lambda \mathbf{x}_1$$

and

$$\mathbf{B}\mathbf{x}_1 + \mathbf{A}\mathbf{x}_2 = \lambda \mathbf{x}_2$$

so that

$$[\mathbf{A} \pm \mathbf{B}][\mathbf{x}_1 \pm \mathbf{x}_2] = \lambda [\mathbf{x}_1 \pm \mathbf{x}_2].$$

Thus λ is a characteristic value of $\mathbf{A} \pm \mathbf{B}$ with characteristic vector $\mathbf{x}_1 \pm \mathbf{x}_2$.

What happens if $x_1 = \pm x_2$? This proof does not work. We can, however, proceed as follows: Since

$$-\frac{1}{2}\begin{bmatrix} -\mathbf{I} & -\mathbf{I} \\ -\mathbf{I} & \mathbf{I} \end{bmatrix}\begin{bmatrix} \mathbf{A} & \mathbf{B} \\ \mathbf{B} & \mathbf{A} \end{bmatrix}\begin{bmatrix} \mathbf{I} & \mathbf{I} \\ \mathbf{I} & -\mathbf{I} \end{bmatrix} = \begin{bmatrix} \mathbf{A}+\mathbf{B} & 0 \\ 0 & \mathbf{A}-\mathbf{B} \end{bmatrix}$$

the matrix

$$\mathcal{A} = \begin{bmatrix} \mathbf{A} & \mathbf{B} \\ \mathbf{B} & \mathbf{A} \end{bmatrix} \quad \text{and} \quad \mathcal{B} = \begin{bmatrix} \mathbf{A}+\mathbf{B} & 0 \\ 0 & \mathbf{A}-\mathbf{B} \end{bmatrix}$$

are similar and have the same characteristic values. Thus the characteristic values of $\mathcal{A}$ are those of $\mathbf{A}+\mathbf{B}$ and of $\mathbf{A}-\mathbf{B}$. The characteristic vectors of $\mathcal{B}$ are $[\mathbf{y}, \mathbf{z}]'$ where $\mathbf{y}, \mathbf{z}$ are the characteristic vectors of $\mathbf{A}+\mathbf{B}$, $\mathbf{A}-\mathbf{B}$ and those of $\mathcal{A}$ are

$$\begin{bmatrix} \mathbf{I} & \mathbf{I} \\ \mathbf{I} & -\mathbf{I} \end{bmatrix}\begin{bmatrix} y \\ z \end{bmatrix} = \begin{bmatrix} y+z \\ y-z \end{bmatrix}.$$

6.7. Solution

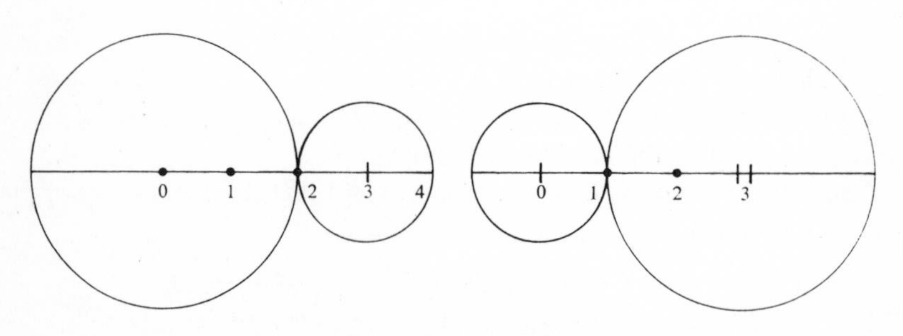

Characteristic values: 1,2

$$\mathbf{A}_1 = \begin{bmatrix} 0 & 1 \\ -2 & 3 \end{bmatrix}.$$

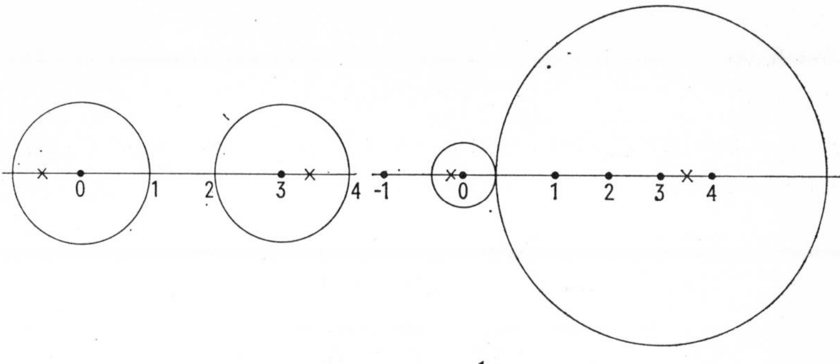

characteristic values: $\frac{1}{2}(3 \pm \sqrt{13})$

$$B_1 = \begin{bmatrix} 0 & (3-\sqrt{5})/2 \\ (3+\sqrt{5})/2 & 3 \end{bmatrix}.$$

REMARKS

(1) There is a considerable body of results known about the Gerschgorin circles. The above simple examples illustrate some of it. For instance, in "general" a characteristic value can lie on the circumference of a Gerschgorin circle only if it lies on the circumferences of all the circles. (The exceptional cases are when the matrix is "reducible", e.g., in the case of the matrix $\begin{bmatrix} 1 & 1 \\ 0 & 3 \end{bmatrix}$ the characteristic value 3 is on the circumference of the point circle center 3 but not on the other circumference.)

(2) If the Gerschgorin region Ω breaks up into separate domains, then there are the "natural" number of characteristic values in each domain.

An extreme case is that of a diagonal matrix with distinct elements — here there is exactly one characteristic value in each point-circle.

(3) If we apply a diagonal similarity to a matrix $\boldsymbol{A}$, i.e. if we consider

$$\boldsymbol{A}_1 = \boldsymbol{D}^{-1} \boldsymbol{A} \boldsymbol{D}$$

we note that the diagonal elements of $\boldsymbol{A}_1$ are the same as those of $\boldsymbol{A}$ so that the centers of the Gerschgorin circles for $\boldsymbol{A}_1$ are the same as those for $\boldsymbol{A}$. However the radii may change and judicious choice of $\boldsymbol{D}$ may result in more information about the location of particular characteristic values, especially isolated ones. This is illustrated in $\boldsymbol{B}$ where we have shrunk the circle center 0 as much as possible, letting the circle center 3 expand. This diagram, combined with preceding remark 2, assures us that there is a characteristic value within the smaller circle.

6.8. Solution

We have

$$\text{(1)} \qquad \det \mathbf{A} = \det \mathbf{A}_1$$

where $\mathbf{A}_1$ is the matrix got from $\mathbf{A}$ by the first major step in the Gauss transformation, i.e.

$$\text{row}_1 \mathbf{A}_1 = \text{row}_1 \mathbf{A}$$

$$\text{row}_i \mathbf{A}_1 = \text{row}_i \mathbf{A} - (a_{i1}/a_{11}) \text{ row}_1 \mathbf{A}, \quad i = 2, 3, \ldots, n.$$

Clearly

$$\text{(2)} \qquad \det \mathbf{A}_1 = a_{11} \det \mathbf{A}_1^{(1)}$$

where $\mathbf{A}_1^{(1)}$ is the matrix of order $n-1$ indicated:

$$\mathbf{A}_1 = \begin{bmatrix} a_{11} & \mathbf{a}' \\ 0 & \mathbf{A}_1^{(1)} \end{bmatrix}.$$

We shall show that $\mathbf{A}_1^{(1)}$ has a dominant diagonal. Looking at the i-th row of $\mathbf{A}_1^{(1)}$ we find

$$\sum_{\substack{k=2\\k\neq i}}^{n} \left| a_{ik} - \frac{a_{i1}}{a_{11}} a_{1k} \right| \leqq \sum_{\substack{k=2\\k\neq i}}^{n} |a_{ik}| + \sum_{\substack{k=2\\k\neq i}}^{n} |a_{i1} a_{1k}/a_{11}|$$

$$= \sum_{\substack{k=2\\k\neq i}}^{n} |a_{ik}| + |a_{i1}| \sum_{k=2}^{n} |(a_{1k}/a_{11})| - |a_{i1} a_{1i}|/|a_{11}|$$

$$\leqq \sum_{\substack{k=2\\k\neq i}}^{n} |a_{ik}| + |a_{i1}| - |a_{i1} a_{1i}|/|a_{11}|$$

(since the first row of $\mathbf{A}$ is strictly diagonally dominant; note that we cannot put a strict inequality here for a_{i1} might be zero)

$$= \sum_{\substack{k=1\\k\neq i}}^{n} |a_{ik}| - |a_{i1} a_{1i}/a_{11}|$$

$$< |a_{ii}| - |a_{i1} a_{1i}/a_{11}|$$

$$= [\mathbf{A}_1^{(1)}]_{ii}$$

since the i-th row of $\mathbf{A}$ is strictly diagonally dominant.

The proof is completed by induction. It is clear that a strictly dominant diagonal matrix of order 1 is non-singular. Assuming that strictly diagonally dominant matrices of order $n-1$ are non-singular it follows that $\mathbf{A}_1^{(1)}$ is non-singular and, by (1) and (2), so is $\mathbf{A}$.

6.9. Solution

From the Rayleigh Quotient Theorem (Problem 1.9) we have

$$\varrho(\mathbf{A}) = \max_{x'x=1} x'\mathbf{A}x.$$

This implies, taking $x = e_i$, that

$$\varrho(\mathbf{A}) \geqq a_{ii}.$$

Here we use the fact that corresponding to the characteristic value ϱ there is a positive characteristic vector $x = [x_1, x_2, \ldots, x_n]'$. Hence from $\mathbf{A}x = \varrho x$ it follows that

$$(a_{i1} - \varrho)x_1 + a_{12}x_2 + \ldots + a_{1n}x_n = 0$$

and, *unless* $n=1$, $a_{11} - \varrho$ must be negative; similarly for $i = 2, 3, \ldots, n$.

6.10. Solution

The characteristic vectors are

$$\begin{bmatrix} 2 \\ 1 \\ \sqrt{3} \end{bmatrix}, \quad \begin{bmatrix} -2 \\ 1 \\ \sqrt{3} \end{bmatrix}, \quad \begin{bmatrix} 0 \\ -\sqrt{3} \\ 1 \end{bmatrix}.$$

6.11. Solution

The whole solution to the characteristic value problem can be written as

$$\mathbf{AM} = \mathbf{M}\Lambda \tag{1}$$

where $\Lambda = \operatorname{diag}[\alpha_1, \ldots, \alpha_n]$. Multiplying (1) by $\mathbf{A}$ we get, using (1),

$$\mathbf{A}^2\mathbf{M} = \mathbf{AM}\Lambda = \mathbf{M}\Lambda^2. \tag{2}$$

Transposing (1) we find

$$\mathbf{M}'\mathbf{A} = \Lambda\mathbf{M}' \tag{3}$$

Multiplying (1) and (3) gives

$$\mathbf{M}'\mathbf{A}^2\mathbf{M} = \Lambda\mathbf{M}'\mathbf{M}\Lambda,$$

which, using (2), can be written as

$$\mathbf{M}'\mathbf{M}\Lambda^2 = \Lambda\mathbf{M}'\mathbf{M}\Lambda,$$

and since Λ is non-singular

$$\mathbf{M}'\mathbf{M} = \Lambda\mathbf{M}'\mathbf{M}\Lambda^{-1}.$$

We now recall the effect of a diagonal similarity (Problem 4.14) which shows that

$$[\mathbf{M}'\mathbf{M}]_{ij} = [\mathbf{M}'\mathbf{M}]_{ij}\,\alpha_i/\alpha_j.$$

Since the α's are distinct this implies

$$[\mathbf{M}'\mathbf{M}]_{ij} = 0 \quad \text{if} \quad i \neq j,$$

i.e. $\boldsymbol{M}'\boldsymbol{M}$ is diagonal.

6.12. Solution

The characteristic values of $\boldsymbol{AB}$ consist of those of $\boldsymbol{BA}$, together with $n-m$ zeros. (See, e.g., Faddeev & Sominskii.)

The characteristic polynomial of $\boldsymbol{AB}$ is

$$-\lambda^3+18\lambda^2-81\lambda\equiv-\lambda(\lambda-9)^2.$$

It follows from the preceding remark that the characteristic polynomial of $\boldsymbol{BA}$ must be $(\lambda-9)^2$. One way to establish the given expression for $\boldsymbol{BA}$ is to observe that

$$(\boldsymbol{AB})^2=9\boldsymbol{AB}$$

and hence

$$\begin{aligned}(\boldsymbol{BA})^3&=\boldsymbol{B}(\boldsymbol{AB})^2\boldsymbol{A}\\&=\boldsymbol{B}(9\boldsymbol{AB})\boldsymbol{A}\\&=9(\boldsymbol{BA})^2.\end{aligned}$$

Since $\boldsymbol{BA}$ is non-singular it follows that $\boldsymbol{BA}=9\boldsymbol{I}$.

The following proof of the result quoted is due to Householder (1972).

We use the Schur complement (Problem 4.10). Let $\boldsymbol{A}$ be an $m\times n$ matrix and $\boldsymbol{B}$ and $n\times m$ matrix. Suppose $m\leqq n$. Consider the matrix

$$\mathcal{A}=\begin{bmatrix}\lambda\boldsymbol{I}_m & \boldsymbol{A}\\ \boldsymbol{B} & \lambda\boldsymbol{I}_n\end{bmatrix}.$$

Then the results of the problem mentioned give

$$\det\mathcal{A}=\det\lambda\boldsymbol{I}_m\det(\lambda\boldsymbol{I}_n-\boldsymbol{B}\lambda^{-1}\boldsymbol{A})=\lambda^{m-n}\det(\lambda^2\boldsymbol{I}-\boldsymbol{BA})$$

and

$$\det\mathcal{A}=\det\lambda\boldsymbol{I}_n\det(\lambda\boldsymbol{I}_m-\boldsymbol{A}\lambda^{-1}\boldsymbol{B})=\lambda^{n-m}\det(\lambda^2\boldsymbol{I}-\boldsymbol{AB}).$$

Hence, writing $\lambda^2=\mu$ and equating the two expressions for $\det\mathcal{A}$ we get

$$\det(\mu\boldsymbol{I}-\boldsymbol{BA})=\mu^{n-m}\det(\mu\boldsymbol{I}-\boldsymbol{AB})$$

which we can interpret as follows: The characteristic roots of $\boldsymbol{BA}$ are those of $\boldsymbol{AB}$, together with $n-m$ zeros.

6.13. Solution

No, for if it were $\boldsymbol{Q}=\boldsymbol{x}'\boldsymbol{A}\boldsymbol{x}$ would be positive for $\boldsymbol{x}'=[0, 0, 1]$ and it vanishes for this vector.

Alternatively, we can easily calculate

$$\det[\boldsymbol{A}-\lambda\boldsymbol{I}]=(2-\lambda)(\lambda+1)^2$$

so that the characteristic values of $\boldsymbol{A}$, which are 2, -1, -1, are not all positive, again showing that $\boldsymbol{A}$ is not positive definite.

The standard Lagrangian method of reduction of a quadratic form to a sum of squares by "eliminating" one variable at a time does not apply here since the diagonal terms all vanish. We therefore use the special device of transforming from

$$x_1, x_2, x_3 \quad \text{to} \quad \xi_1, \xi_2, \xi_3 \quad \text{by putting} \quad x_1=\xi_1, \quad x_2=\xi_1+\xi_2, \quad x_3=\xi_3$$

so that $\xi_1=x_1$, $\xi_2=x_2-x_1$, $\xi_3=x_3$. We find

$$Q(x)\to\mathcal{Q}(\xi)\equiv(\xi_1+\xi_3)(\xi_1+\xi_2)+\xi_1\xi_3$$
$$=\xi_1^2+\xi_1\xi_2+2\xi_1\xi_3+\xi_2\xi_3$$

and we can apply the standard method to this. Indeed

$$\mathcal{Q}(\xi)=\xi_1^2+2\xi_1(\tfrac{1}{2}\xi_2+\xi_3)+(\tfrac{1}{2}\xi_2+\xi_3)^2-(\tfrac{1}{2}\xi_2+\xi_3)^2+\xi_2\xi_3$$
$$=(\xi_1+\tfrac{1}{2}\xi_2+\xi_3)^2-\tfrac{1}{4}\xi_2^2-\xi_3^2$$

and we can now change back to get

$$Q(x)\equiv(\tfrac{1}{2}x_1+\tfrac{1}{2}x_2+x_3)^2-(\tfrac{1}{2}x_2-\tfrac{1}{2}x_1)^2-x_3^2$$

which we can check by expanding.

6.14. Solution

See p. 101—3.

6.15. Solution

Consider

$$\mathcal{A}(\lambda)=\begin{bmatrix}-\lambda\boldsymbol{I} & \boldsymbol{A}^*\\ \boldsymbol{A} & -\lambda\boldsymbol{I}\end{bmatrix}.$$

By "block Gaussian elimination"

$$\begin{bmatrix}\boldsymbol{I} & 0\\ \lambda^{-1}\boldsymbol{A} & \boldsymbol{I}\end{bmatrix}\begin{bmatrix}-\lambda\boldsymbol{I} & \boldsymbol{A}^*\\ \boldsymbol{A} & -\lambda\boldsymbol{I}\end{bmatrix}=\begin{bmatrix}-\lambda\boldsymbol{I} & \boldsymbol{A}^*\\ 0 & \lambda^{-1}\boldsymbol{A}\boldsymbol{A}^*-\lambda\boldsymbol{I}\end{bmatrix}.$$

Hence

$$\det\mathcal{A}(\lambda)=\det(-\lambda\boldsymbol{I})\det(\lambda^{-1}\boldsymbol{A}\boldsymbol{A}^*-\lambda\boldsymbol{I})$$
$$=-\lambda^n\det(\lambda^{-1}\boldsymbol{A}\boldsymbol{A}^*-\lambda\boldsymbol{I})$$
$$=\det(\lambda^2\boldsymbol{I}-\boldsymbol{A}\boldsymbol{A}^*).$$

Thus the characteristic values are the positive and negative square roots of the characteristic values of $\boldsymbol{A}\boldsymbol{A}^*$, i.e. the singular values of $\boldsymbol{A}$ and their negatives.

6.16. Solution

See J. Williamson, Bull. American Math. Soc., *37*, 585—590 (1931). We require the following result of Schur:

If $\mathbf{A}$ *is any (complex) matrix, there is a unitary matrix* $\mathbf{U}$ *such that* $\mathbf{U}^*\mathbf{A}\mathbf{U}$ *is a triangular matrix.*

Proof. The result is trivial when $\mathbf{A}$ is 1×1. Assume that the result has been established for $(n-1)\times(n-1)$ matrices. Let $\boldsymbol{a}$ be a characteristic vector of $\mathbf{A}$ corresponding to the characteristic value α. Choose (as in Problem 4.24) other vectors $\boldsymbol{x}_2, \ldots, \boldsymbol{x}_n$ so that

$$\mathbf{V}=[\boldsymbol{a}, \boldsymbol{x}_2, \ldots, \boldsymbol{x}_n]$$

is unitary. Consider

$$\mathbf{V}^*\mathbf{A}\mathbf{V}=\begin{bmatrix}\boldsymbol{a}^*\\ \boldsymbol{x}_2^*\\ \vdots\\ \boldsymbol{x}_n^*\end{bmatrix}\mathbf{A}[\boldsymbol{a}, \boldsymbol{x}_2, \ldots, \boldsymbol{x}_n]=\begin{bmatrix}\boldsymbol{a}^*\\ \boldsymbol{x}_2^*\\ \vdots\\ \boldsymbol{x}_n^*\end{bmatrix}[\alpha\boldsymbol{a}, \mathbf{A}\boldsymbol{x}_2, \ldots, \mathbf{A}\boldsymbol{x}_n]$$

$$=\left[\begin{array}{c|c}\alpha & \hat{\boldsymbol{x}}^*\\ \hline 0 & \\ \vdots & \mathbf{A}_1\\ 0 & \end{array}\right], \quad \text{say.}$$

The unitary similarity does not change the characteristic roots so that those of $\hat{\mathbf{A}}_1$ are the remaining characteristic roots of $\mathbf{A}$. By our inductive hypothesis there is an $(n-1)\times(n-1)$ unitary matrix $\hat{\mathbf{U}}$ such that

$$\hat{\mathbf{U}}^*\hat{\mathbf{A}}_1\hat{\mathbf{U}}=\hat{\mathbf{R}}, \quad \text{an upper triangular matrix.}$$

Hence

$$\begin{bmatrix}1 & 0\\ 0 & \hat{\mathbf{U}}^*\end{bmatrix}\mathbf{V}^*\mathbf{A}\mathbf{V}\begin{bmatrix}1 & 0\\ 0 & \hat{\mathbf{U}}\end{bmatrix}=\begin{bmatrix}1 & 0\\ 0 & \hat{\mathbf{U}}\end{bmatrix}\begin{bmatrix}\alpha & \hat{\boldsymbol{x}}^*\\ 0 & \hat{\mathbf{A}}_1\end{bmatrix}\begin{bmatrix}1 & 0\\ 0 & \hat{\mathbf{U}}\end{bmatrix}=\begin{bmatrix}\alpha & \hat{\boldsymbol{x}}^*\\ 0 & \hat{\mathbf{R}}\end{bmatrix},$$

a triangular matrix. Since $\mathbf{V}\begin{bmatrix}1 & 0\\ 0 & \hat{\mathbf{U}}\end{bmatrix}$ is unitary the proof is complete.

An easy consequence of this result is the fact that any normal matrix $\mathbf{A}$ (i.e. satisfying $\mathbf{A}\mathbf{A}^*=\mathbf{A}^*\mathbf{A}$) is unitarily similar to a diagonal matrix.

We outline the proof of Williamson's Theorem in the case of 2×2 block matrices of 2×2 matrices. It will be convenient to change the notation so that

$$\mathcal{A}=\begin{bmatrix}\mathbf{A} & \mathbf{B}\\ \mathbf{C} & \mathbf{D}\end{bmatrix}$$

where $\mathbf{A}=a(\mathbf{M})$, $\mathbf{B}=b(\mathbf{M})$, ... with $a, b, \ldots$ polynomials. By the result of Schur just established there is a unitary matrix $\mathbf{U}$ which triangularizes $\mathbf{M}$, i.e. $\mathbf{U}^*\mathbf{M}\mathbf{U}=\mathbf{R}$; it is easy to see that $\mathbf{U}$ also triangularizes $\mathbf{A}, \mathbf{B}, \ldots$ so that

$\boldsymbol{U}^* \boldsymbol{A} \boldsymbol{U} = a(\mathbf{R})$, $\boldsymbol{U}^* \boldsymbol{B} \boldsymbol{U} = b(\mathbf{R})$, ... Hence, if $\boldsymbol{M}$ has characteristic roots μ, ν:

$$\begin{bmatrix} \boldsymbol{U}^* & 0 \\ 0 & \boldsymbol{U}^* \end{bmatrix} \mathcal{A} \begin{bmatrix} \boldsymbol{U} & 0 \\ 0 & \boldsymbol{U} \end{bmatrix} = \begin{bmatrix} a(\mathbf{R}) & b(\mathbf{R}) \\ c(\mathbf{R}) & d(\mathbf{R}) \end{bmatrix} = \begin{bmatrix} a(\mu) & * & b(\mu) & * \\ 0 & a(\nu) & 0 & b(\nu) \\ c(\mu) & * & d(\mu) & * \\ 0 & c(\nu) & 0 & d(\nu) \end{bmatrix}$$

where the elements marked $*$ are irrelevant in the present context.

Since we have performed a similarity on $\mathcal{A}$, the characteristic values of $\mathcal{A}$ are those of the matrix on the right and therefore of the matrix

$$\mathcal{A}_1 = \begin{bmatrix} a(\mu) & b(\mu) & * & * \\ c(\mu) & d(\mu) & * & * \\ 0 & 0 & a(\nu) & b(\nu) \\ 0 & 0 & c(\nu) & d(\nu) \end{bmatrix}$$

obtained by submitting that matrix to a similarity which interchanges the second and third rows and the second and third columns. But the characteristic values of $\mathcal{A}_1$ are those of

$$\begin{bmatrix} a(\mu) & b(\mu) \\ c(\mu) & d(\mu) \end{bmatrix} \quad \text{together with those of} \quad \begin{bmatrix} a(\nu) & b(\nu) \\ c(\nu) & d(\nu) \end{bmatrix}$$

which is Williamson's Theorem in the special case.

6.17. Solution

Compare Problem 6.2. $\boldsymbol{A}$ is the sum of two matrices $[c_i c_j]$ and $[r_i r_j]$ each of rank $\leqq 1$. Hence rank $\boldsymbol{A} \leqq 2$.

We use the following general theorem:

$$\det(\boldsymbol{A} - \lambda \boldsymbol{I}) \equiv (-1)^n [\lambda^n - \gamma_1 \lambda^{n-1} + \gamma_2 \lambda^{n-2} \ldots + (-1)^n \gamma_n]$$

where for $r = 1, 2, \ldots, n$, γ_r is the sum of all the principal minors of order r of $\boldsymbol{A}$.

Clearly

$$\gamma_1 = \operatorname{trace} A = \sum c_i^2 + \sum r_i^2$$

and

$$\begin{aligned} \gamma_2 = \sum_{\substack{i<j \\ i,\, j=1}}^{n} (a_{ii} a_{jj} - a_{ij}^2) &= \sum_{\substack{i<j \\ i,\, j=1}}^{n} [(c_i^2 + r_i^2)(c_j^2 + r_j^2) - (c_i c_j + r_i r_j)^2] \\ &= \sum (c_i r_j - r_i c_j)^2 \\ &= \sum c_i^2 \sum r_j^2 - (\sum c_i r_i)^2. \end{aligned}$$

Hence the characteristic polynomial is

$$(-1)^n \{\lambda^n - (\|\boldsymbol{c}\|^2 + \|\boldsymbol{r}\|^2) \lambda^{n-1} + (\|\boldsymbol{c}\|^2 \|\boldsymbol{r}\|^2 - (\boldsymbol{r}, \boldsymbol{c})^2) \lambda^{n-2}\}.$$

6.18. Solution

$\boldsymbol{A}$ is *not* symmetric and so cannot be orthogonally similar to a diagonal matrix. Since

$$\det(\boldsymbol{A}-\lambda\boldsymbol{I}) = -(\lambda-1)(\lambda^2+\tfrac{2}{3}\lambda+1)$$

the characteristic values are $\alpha_1=1$ and $\alpha_{2,3}=(-1\pm 2\sqrt{2}i)/3$, all of absolute value 1. Corresponding characteristic vectors are

$$\boldsymbol{v}_1=\frac{1}{\sqrt{2}}\begin{bmatrix}1\\-1\\0\end{bmatrix},\quad \boldsymbol{v}_2=\frac{1}{2}\begin{bmatrix}1\\1\\\sqrt{2}i\end{bmatrix},\quad \boldsymbol{v}_3=\frac{1}{2}\begin{bmatrix}1\\1\\-\sqrt{2}i\end{bmatrix}.$$

Clearly if $\boldsymbol{U}=[\boldsymbol{v}_1, \boldsymbol{v}_2, \boldsymbol{v}_3]$ then $\boldsymbol{U}$ is unitary and $\boldsymbol{U}^*\boldsymbol{A}\boldsymbol{U}=\operatorname{diag}[1, \alpha_2, \alpha_3]$.

Generally, if $\boldsymbol{A}$ is normal, i.e. if $\boldsymbol{A}\boldsymbol{A}^*=\boldsymbol{A}^*\boldsymbol{A}$, then $\boldsymbol{A}$ is unitarily similar to a diagonal matrix. (Cf. Problem 6.16.)

6.19. Solution

Since $\boldsymbol{x}\boldsymbol{x}^*=\boldsymbol{x}^*\boldsymbol{x}$ the first relation is trivial. To deal with the second use the fact that there is a unitary $\boldsymbol{S}$ such that

$$\boldsymbol{S}^{-1}\boldsymbol{A}\boldsymbol{S}=\operatorname{diag}[\lambda_1, \lambda_2, \ldots, \lambda_n].$$

The characteristic vector of $\boldsymbol{A}$ corresponding to λ_i is $\boldsymbol{u}_i=\boldsymbol{S}\boldsymbol{e}_i$, $i=1, 2, \ldots, n$. Let $\mu_i=\sqrt{\lambda_i}$, $i=1, 2, \ldots, n$, the positive square root being taken in all cases. Define

$$\boldsymbol{B}=\boldsymbol{S}\operatorname{diag}[\mu_1, \mu_2, \ldots, \mu_n]\boldsymbol{S}^{-1}.$$

Then it is clear that $\boldsymbol{u}_i$ is also a characteristic vector of $\boldsymbol{B}$, corresponding to the characteristic value μ_i. We put $\alpha=\max\mu_i$ and $\beta^{-1}=\min\mu_i$. Consider the matrix

$$\begin{aligned}(1)\qquad \boldsymbol{R}&=(\beta\boldsymbol{B}-\beta^{-1}\boldsymbol{B}^{-1})(\alpha\boldsymbol{B}^{-1}-\alpha^{-1}\boldsymbol{B})\\ &=(\alpha\beta+\alpha^{-1}\beta^{-1})\boldsymbol{I}-\alpha\beta^{-1}\boldsymbol{A}^{-1}-\alpha^{-1}\beta\boldsymbol{A}.\end{aligned}$$

The characteristic vectors of $\boldsymbol{R}$ are the $\boldsymbol{u}_i$ and the corresponding characteristic values are $\alpha^{-1}\beta(1-\alpha_i^{-1}\beta^{-2})(1-\alpha_i\alpha^{-2})\geqq 0$. Hence $\boldsymbol{R}$, being hermitian, is positive semi-definite.

Take any $\boldsymbol{x}\neq 0$. Express it as $\sum\gamma_i\boldsymbol{u}_i$. Then, assuming the $\boldsymbol{u}_i$ normalised.

$$\boldsymbol{x}^*\boldsymbol{R}\boldsymbol{x}=\sum|\gamma_i|^2>0.$$

Using the relation (1) we conclude that

$$(\alpha\beta+\alpha^{-1}\beta^{-1})\boldsymbol{x}^*\boldsymbol{x}\geqq\alpha\beta^{-1}\boldsymbol{x}^*\boldsymbol{A}^{-1}\boldsymbol{x}+\alpha^{-1}\beta\boldsymbol{x}^*\boldsymbol{A}\boldsymbol{x}$$

and the arithmetic-geometric mean theorem tells us that the right-hand side is not less than

$$2\sqrt{(x^* \mathbf{A} x)(x^* \mathbf{A}^{-1} x)}.$$

Rearranging and squaring we have

$$\frac{(\alpha\beta+\alpha^{-1}\beta^{-1})^2}{4} \geqq \frac{(x^* \mathbf{A} x)(x^* \mathbf{A}^{-1} x)}{(x^* x)^2}$$

as required.

If we use the euclidean vector norm $\|\cdot\|_2$ then the induced matrix norms of $\mathbf{A}$, $\mathbf{A}^{-1}$ are clearly

$$\|\mathbf{A}\| = \alpha, \quad \|\mathbf{A}^{-1}\| = \beta$$

and the result can be expressed in the form

$$\frac{(x^* \mathbf{A} x)(x^* \mathbf{A}^{-1} x)}{(x^* x)^2} \leqq \frac{(\varkappa^{1/2}(\mathbf{A}) + \varkappa^{-1/2}(\mathbf{A}))^2}{4}.$$

In the special case when $\mathbf{A}$ is diagonal, if we write $m = \min_i \lambda_i$, $M = \max_i \lambda_i$, the inequality becomes

$$\frac{(\sum \lambda_i |x_i|^2)(\sum \lambda_i^{-1} |x_i|^2)}{(\sum |x_i|^2)^2} \leqq \frac{(m+M)^2}{4mM}$$

or, replacing x by a unit vector $\tilde{x}$, and writing q_i for $|x_i^2|$ so that $\sum q_i = 1$,

$$(\sum \lambda_i q_i)(\sum \lambda_i^{-1} q_i) \leqq \frac{(m+M)^2}{4mM}.$$

Chapter 7

7.1. Solution

$$\mu^{(3)} = 16.9850, \quad v^{(3)} = [1,\ 9.9801,\ 5.0017]'$$

$$\mu^{(4)} = 15.7843, \quad v^{(4)} = [1,\ 9.9807,\ 4.9928]'$$

$$\mu^{(5)} = 15.7731, \quad v^{(5)} = [1,\ 9.9947,\ 4.9979]'$$

$$\mu^{(6)} = 15.7925, \quad v^{(6)} = [1,\ 9.9987,\ 4.9995]'$$

The exact results are

$$\lambda_1 = 15.8, \quad v_1 = [1,\ 10,\ 5]'.$$

The other characteristic pairs are

$$\lambda_2 = 3.16, \quad v_2 = [3,\ 4,\ 5]'$$

$$\lambda_3 = 1.58, \quad v_3 = [2,\ 1.6,\ 1]'.$$

7.2. Solution

The characteristic pairs are

$$45,\ [-3,\ -1,\ 2]'$$

$$2,\ [-3,\ -2,\ 3]'$$

$$1,\ [-2,\ -1,\ 2]'.$$

7.3. Solution

For details see E. Bodewig, Math. Tables Aids Comp. *8*, 237—239 (1954). Convergence is very slow. After 1200 iterations the dominant characteristic value is only good to about 4D. There are two reasons:

(a) the characteristic values are not well separated

$$-8.02857\,835, \quad 7.93290\,472, \quad 5.66886\,436, \quad -1.57319\,074$$

and

(b) the dominant characteristic vector

$$[1,\ \ 2.50146030,\ \ -.75773064,\ \ -2.56421\,169]'$$

is very nearly orthogonal to the chosen initial guess

$$[1,\ 1,\ 1,\ 1]'$$

— in fact their inner product is

$$.17951\,797$$

so that the cosine of the angle between them is

$$.17951\ 797 \div \sqrt{1^2+1^2+1^2+1^2} \times \sqrt{1^2+(2.50\ldots)^2+(.75\ldots)^2+(2.56\ldots)^2}$$

which is about $.0310 \doteqdot \cos 88°$.

7.4. Solution

With the notation of the text we find

$$\boldsymbol{H}_4 \boldsymbol{v}^{(0)} = [1.508333,\ .860000,\ .613333,\ .479524]'$$

so that

$$\mu^{(1)} = 1.508333, \quad \boldsymbol{v}^{(1)} = [1,\ .570166,\ .406630,\ .317917]'.$$

Compare this with the 10 D results given on p. 162.

7.5. Solution

We have to assume that the moduli of the roots are different so that $\boldsymbol{A}$ has necessarily distinct characteristic roots, λ, μ say, where $|\lambda| > |\mu|$. Hence $\boldsymbol{A}$ is diagonalizable. We denote by $[a, c]'$ and $[b, d]'$ the characteristic vectors of $\mathbf{A}$. Then, since these are distinct, we may assume $ad-bc=1$. This means that if

$$\mathbf{T}=\begin{bmatrix} a & b \\ c & d \end{bmatrix} \quad \text{then} \quad \mathbf{T}^{-1}=\begin{bmatrix} d & -b \\ -c & a \end{bmatrix} \quad \text{and} \quad \mathbf{T}^{-1}\mathbf{A}\mathbf{T}=\begin{bmatrix} \lambda & 0 \\ 0 & \mu \end{bmatrix}=\Lambda, \quad \text{say.}$$

To compute $\mathbf{A}^n$ we observe that since $\mathbf{A}=\mathbf{T}\Lambda\mathbf{T}^{-1}$ we have $\mathbf{A}^n= =\mathbf{T}\Lambda\mathbf{T}^{-1}\cdot\mathbf{T}\Lambda\mathbf{T}^{-1}\cdot\ldots\cdot\mathbf{T}\Lambda\mathbf{T}^{-1}$ and this product collapses so that

$$\mathbf{A}^n=\mathbf{T}\Lambda^n\mathbf{T}^{-1}=\mathbf{T}\begin{bmatrix} \lambda^n & 0 \\ 0 & \mu^n \end{bmatrix}\mathbf{T}^{-1}=\begin{bmatrix} ad\lambda^n-bc\mu^n, & -ab(\lambda^n-\mu^n) \\ cd(\lambda^n-\mu^n), & -bc\lambda^n+ad\mu^n \end{bmatrix}.$$

Hence

$$\text{(1)} \qquad \mathbf{A}^n\begin{bmatrix} x_1 \\ x_2 \end{bmatrix}=\begin{bmatrix} (ad\lambda^n-bc\mu^n)x_1-ab(\lambda^n-\mu^n)x_2 \\ cd(\lambda^n-\mu^n)x_1-(bc\lambda^n-ad\mu^n)x_2 \end{bmatrix}.$$

Generally, therefore, the direction of this vector will tend to that of the dominant vector

$$\begin{bmatrix} adx_1-abx_2 \\ cdx_1-bcx_2 \end{bmatrix}=(dx_1-bx_2)\begin{bmatrix} a \\ c \end{bmatrix}$$

since $\lambda^n \gg \mu^n$ as $n\to\infty$. Difficulties can arise if $dx_1-bx_2=0$ which can be interpreted as meaning that our initial vector $[x_1, x_2]'$ is parallel to the other characteristic vector of $\mathbf{A}$, containing no component in the direction of the dominant vector $[a, c]'$.

It is also clear from (1) that the ratio of the lengths of successive vectors $\mathbf{A}^n[x_1, x_2]'$ approaches λ.

7.6. Solution

If we take $\alpha=\frac{1}{2}(20+5)$ then the roots of the new matrix $\mathbf{A}-\alpha\mathbf{I}$ are

$$-\frac{23}{2}, \quad -\frac{15}{2}, \quad -\frac{5}{2}, \quad \frac{15}{2}$$

and there is a separation factor of 15/23 which makes the power method quite attractive.

The smallest characteristic values of the matrices in Problems 7.1, 7.2 have been given above. That for $\mathbf{H}_4$ has been given in Problem 5.13 (viii).

7.7. Solution

In the special case we have

$$\mathbf{S}=\begin{bmatrix}1&0&0\\0&1&0\\1&0&1\end{bmatrix},\quad \mathbf{S}^{-1}=\begin{bmatrix}1&0&0\\0&1&0\\-1&0&1\end{bmatrix},\quad \mathbf{S}^{-1}\mathbf{A}\mathbf{S}=\begin{bmatrix}1&1&-4\\0&2&2\\0&0&4\end{bmatrix}.$$

We check that $[1, 1]'$ is a characteristic vector of $\mathbf{A}_1$:

$$\begin{bmatrix}2&2\\0&4\end{bmatrix}\begin{bmatrix}1\\1\end{bmatrix}=4\begin{bmatrix}1\\1\end{bmatrix}.$$

Then

$$\hat{y}=\frac{4-1}{[1,\ -4]\begin{bmatrix}1\\1\end{bmatrix}}\begin{bmatrix}1\\1\end{bmatrix}=\begin{bmatrix}-1\\-1\end{bmatrix}$$

so that a characteristic vector of $\mathbf{S}^{-1}\mathbf{A}\mathbf{S}$ is $[1, -1, -1]'$ and that for $\mathbf{A}$ is

$$\mathbf{S}\,y=\begin{bmatrix}1&0&0\\0&1&0\\1&0&1\end{bmatrix}\begin{bmatrix}1\\-1\\-1\end{bmatrix}=\begin{bmatrix}1\\-1\\0\end{bmatrix}.$$

The characteristic vector of $\mathbf{A}$ corresponding to the characteristic value 2 is $[1, 1, 1]'$.

REMARK. The "exceptional case" $\hat{r}'\tilde{y}=0$ *can* arise, for instance, in the case of the matrix

$$\mathbf{B}=\begin{bmatrix}0&1&-1\\2&2&-2\\-4&1&5\end{bmatrix}$$

(a permutation similarity of the given matrix). Take the characteristic pair

$$1,\ [1,\ 0,\ 1]'.$$

We deflate $\mathbf{B}$ to get

$$\mathbf{T}^{-1}\mathbf{B}\mathbf{T}=\begin{bmatrix}1&1&1\\0&2&-2\\0&0&4\end{bmatrix}\quad\text{where}\quad \mathbf{T}=\begin{bmatrix}1&0&0\\0&1&0\\1&0&1\end{bmatrix}.$$

Choose the characteristic pair

$$4,\ [-1,\ 1]'$$

and we have

$$\hat{r}'\hat{y}=[1,\ 1]\begin{bmatrix}-1\\1\end{bmatrix}=0$$

so that the characteristic pair of $\mathbf{T}^{-1}\,\mathbf{B}\,\mathbf{T}$ is

$$4, \; [0, \; -1, \; 1]$$

and the characteristic vector of $\mathbf{B}$ is

$$\begin{bmatrix} 1 & 0 & 0 \\ 0 & 1 & 0 \\ 1 & 0 & 1 \end{bmatrix} \begin{bmatrix} 0 \\ -1 \\ 1 \end{bmatrix} = \begin{bmatrix} 0 \\ -1 \\ 1 \end{bmatrix}.$$

7.8. Solution

Let $\boldsymbol{r}_3'$ be the third row of $\mathbf{A}$. We have $\mathbf{A}\,\boldsymbol{v}=\lambda_3\,\boldsymbol{v}$ and, in particular, $\boldsymbol{r}_3'\,\boldsymbol{v}=\lambda_3$ since the third component of $\boldsymbol{v}$ is 1. Suppose $\lambda\neq\lambda_3$ is a characteristic value of $\mathbf{A}$ and let the corresponding characteristic vector $\boldsymbol{c}$ also be normalized to have its third component unity so that, as before, $\boldsymbol{r}_3'\boldsymbol{c}=\lambda$.

Consider the matrix

$$\tilde{\mathbf{A}}=\mathbf{A}-\boldsymbol{v}\,\boldsymbol{r}_3'.$$

This matrix has as its third row the zero vector. Thus one of its characteristic values is zero and all the characteristic vectors have their third component zero.

We note that

$$\begin{aligned} \tilde{\mathbf{A}}\,(\boldsymbol{v}-\boldsymbol{c}) &= (\mathbf{A}-\boldsymbol{v}\,\boldsymbol{r}_3')\,(\boldsymbol{v}-\boldsymbol{c}) \\ &= \lambda_3\,\boldsymbol{v}-\lambda\,\boldsymbol{c}-\lambda_3\,\boldsymbol{v}+\lambda\,\boldsymbol{v} \\ &= \lambda\,(\boldsymbol{v}-\boldsymbol{c}) \end{aligned}$$

so that $\boldsymbol{v}-\boldsymbol{c}$ is a characteristic vector of $\tilde{\mathbf{A}}$ with characteristic value λ.

In view of what has been said we can get λ as a characteristic value of the principal 2×2 submatrix of $\tilde{\mathbf{A}}$, which is the "deflation" of $\mathbf{A}$. If $\hat{\boldsymbol{b}}$ is the corresponding 2-dimensional characteristic vector and $\boldsymbol{b}=[\hat{\boldsymbol{b}}', 0]'$ then we can choose $\vartheta\neq0$ so that

$$\boldsymbol{c}=\boldsymbol{v}-\vartheta\,\boldsymbol{b}$$

is a characteristic vector of $\mathbf{A}$ with characteristic value λ, i.e.,

$$\mathbf{A}\,(\boldsymbol{v}-\vartheta\,\boldsymbol{b})=\lambda\,(\boldsymbol{v}-\vartheta\,\boldsymbol{b}). \tag{1}$$

In view of our constructions we have

$$[\mathbf{A}-\boldsymbol{v}\,\boldsymbol{r}_3']\,\boldsymbol{b}=\lambda\,\boldsymbol{b}, \quad \text{i.e.} \quad \mathbf{A}\,\boldsymbol{b}=\boldsymbol{v}\,\boldsymbol{r}_3'\,\boldsymbol{b}-\lambda\,\boldsymbol{b}$$

and (1) will be satisfied if

$$\lambda_3\,\boldsymbol{v}-\vartheta\,(\boldsymbol{r}_3'\,\boldsymbol{b})\,\boldsymbol{v}-\vartheta\,\lambda\,\boldsymbol{b}=\lambda\,\boldsymbol{v}-\vartheta\,\lambda\,\boldsymbol{b}$$

which is the case if

$$\vartheta = \frac{\lambda_3 - \lambda}{\mathbf{r}_3' \mathbf{b}}.$$

(Discuss the situation when $\mathbf{r}_3' \mathbf{b} = 0$.)

We illustrate this with a slightly modified version of Problem 7.2.

$\mathbf{A} = \begin{bmatrix} 133 & 135 & 6 \\ -88 & -90 & -6 \\ 44 & 46 & 5 \end{bmatrix}$ has as a characteristic pair, $\lambda_3 = 45$, $\mathbf{v} = [3, -2, 1]'$ and

$$\tilde{\mathbf{A}} = \mathbf{A} - \begin{bmatrix} 3 \\ -2 \\ 1 \end{bmatrix} [44, 46, 5] = \begin{bmatrix} 1 & -3 & -12 \\ 0 & 2 & 4 \\ 0 & 0 & 0 \end{bmatrix}.$$

The characteristic pairs of the deflated matrix

$$\begin{bmatrix} 1 & -3 \\ 0 & 2 \end{bmatrix} \quad \text{are} \quad 1, [1, 0]' \quad \text{and} \quad 2, [3, -1]'.$$

Let us see how to build up the characteristic vector of $\mathbf{A}$ corresponding to the characteristic value 2. We take $\mathbf{b} = [3, -1, 0]'$ to give

$$\vartheta = \frac{45 - 2}{(44 \times 3) + (46 \times (-1))} = \frac{1}{2}$$

and then

$$\mathbf{c} = \begin{bmatrix} 3 \\ -2 \\ 1 \end{bmatrix} - \frac{1}{2} \begin{bmatrix} 3 \\ -1 \\ 0 \end{bmatrix} = \begin{bmatrix} 3/2 \\ -3/2 \\ 1 \end{bmatrix}$$

so that we find the characteristic pair

$$2, [3, -3, 2]'.$$

7.9. Solution

The characteristic values of the matrix are

$$\alpha_r = 2 - 2 \cos (r\theta), \quad r = 1, 2, 3, 4$$

with characteristic vectors

$$\mathbf{c}_r = [\sin r\theta, \sin 2r\theta, \sin 3r\theta, \sin 4r\theta]', \quad r = 1, 2, 3, 4$$

where $\theta = \pi/5$.

We shall go after the characteristic vector $\mathbf{c}_1$ where

$$\mathbf{c}_1' = [\sin (\pi/5), \sin (2\pi/5), \sin (3\pi/5), \sin (4\pi/5)]$$

which is parallel to

$$[1,\ 2\cos(\pi/5),\ 2\cos(\pi/5),\ 1]=[1,\ \tfrac{1}{2}(\sqrt{5}+1),\ \tfrac{1}{2}(\sqrt{5}+1),\ 1]$$

$$\doteqdot[1,\ 1.6180,\ 1.6180,\ 1].$$

The corresponding characteristic value is

$$\alpha_1=2-2\cos(\pi/5)=4\sin^2(\pi/10)=(3-\sqrt{5})/2\doteqdot 0.3820.$$

We shall take

$$\boldsymbol{v}^{(0)}=[1,\ 1,\ 1,\ 1]',\quad \hat{\alpha}=\cdot 4.$$

We find, solving the system (7.2) for $v^{(1)}$:

$$(\boldsymbol{v}^{(1)})'=[-40,\ -65,\ -65,\ -40]=-40\left[1,\ \frac{13}{8},\ \frac{13}{8},\ 1\right].\quad \left(\frac{13}{8}=1.625\right).$$

We repeat, solving the system (2) for $\boldsymbol{v}^{(2)}$ taking the normalized $\boldsymbol{v}^{(1)}$ on the right and find

$$\boldsymbol{v}^{(2)\prime}=\left[\frac{-445}{8}\ -90,\ -90,\ \frac{-445}{8}\right]=\frac{-445}{8}\left[1,\ \frac{144}{89},\ \frac{144}{89},\ 1\right].$$

$$\left(\frac{144}{89}\doteqdot 1.61798\right).$$

It is interesting to see the relative sizes of the components a_i. We find these by taking the scalar product of $\boldsymbol{v}^{(0)}$ with $\boldsymbol{c}_i$ to get, since the c_i's are orthogonal:

$$\boldsymbol{c}_i'\,\boldsymbol{v}^{(0)}=a_i\,(\boldsymbol{c}_i'\,\boldsymbol{c}_i).$$

The scalar product on the right can be evaluated by elementary trigonometry and we find

$$a_1=\frac{10+2\sqrt{5}}{10},\quad a_2=0,\quad a_3=\frac{10-2\sqrt{5}}{10},\quad a_4=0.$$

Question. What is the dominant characteristic value of $\mathbf{C}_4$? What would happen if you would try to find it using the power method, starting with the vector $\boldsymbol{v}^{(0)}$ above? Observe that $\alpha_4=\frac{1}{2}(5+\sqrt{5})$ and that $\boldsymbol{c}_4$ is orthogonal to $\boldsymbol{v}^{(0)}$ as $\boldsymbol{c}_4=[1,\ -(\sqrt{5}+1)/2,\ (\sqrt{5}+1)/2,\ -1]$.

7.10. Solution

G. E. Forsythe (MTAC, *6,* 9—17, esp. 15 (1952).)

The characteristic pairs are

$$
\begin{array}{lrrrr}
30.29, & [\ \ .380, & .526, & .552, & .521]'; \\
3.858, & [\ \ .396, & .614, & -.271, & -.625]'; \\
.8431, & [\ \ .094, & -.302, & .761, & -.568]'; \\
.01015, & [-.830, & .501, & .208, & -.124]'.
\end{array}
$$

Chapter 8

8.1. Solution

The basic equation is

$$\begin{bmatrix} c & s \\ -s & c \end{bmatrix} \begin{bmatrix} A & H \\ H & B \end{bmatrix} \begin{bmatrix} c & -s \\ s & c \end{bmatrix} = \begin{bmatrix} a & 0 \\ 0 & b \end{bmatrix}$$

where c, s are the cosine and sine of the angle θ defined by

$$\tan 2\theta = 2H/(A-B),$$

and the values of a, b are given by

$$a = Ac^2 + Bs^2 + 2Hsc, \quad b = As^2 + Bc^2 - 2Hsc.$$

(Observe that $a+b=A+B$ and that $a^2+b^2=A^2+B^2+2H^2$.)

If we write $n=2H$, $d=A-B$ it follows from elementary trigonometry that

$$
\begin{aligned}
2c^2 &= \{(n^2+d^2)^{1/2}+d\}\,(n^2+d^2)^{-1/2} \\
2s^2 &= \{(n^2+d^2)^{1/2}-d\}\,(n^2+d^2)^{-1/2}, \qquad (1)\\
2sc &= n\,(n^2+d^2)^{-1/2}.
\end{aligned}
$$

If we take case $A=2$, $B=5$, $H=-3$ we find

$$\tan 2\theta = 2/1, \quad n=2, \quad d=1, \quad \text{say.}$$

Hence

$$2c^2 = \{\sqrt{5}+1\}\sqrt{5}, \quad 2s = \{\sqrt{5}-1\}/\sqrt{5}, \quad 2sc = 2/\sqrt{5},$$

which gives, in particular,

$$a = (7-3\sqrt{5})/2, \quad b = (7+3\sqrt{5})/2,$$

results which check with the fact that a, b are the characteristic values of the matrix $\begin{bmatrix} A & H \\ H & B \end{bmatrix}$.

Care must be taken with the ambiguities in c, s, e.g., by taking $\frac{-\pi}{2} \leqq 2\theta \leqq \frac{\pi}{2}$ and then taking c positive and s to have the sign of $\tan 2\theta$. Further, the formulas (1) are numerically unsatisfactory if $n \ll d$ and special tricks must be used.

Since $\arctan 2 = 1.1071$ the required angle of rotation of the axes is $\theta = .5536 \sim 31.72°$.

8.3. Solution

See Solution 8.7 below.

8.6. Solution

The first polynomial has all its roots real and located as follows: $(-11, -10)$, $(-4, -3)$, $(-2, -1)$, two in $(1, 2)$, $(4, 5)$.

The second polynomial has exactly two real zeros, one in the interval $(-2, -1)$, the other in $(6, 7)$.

8.7. Solution

We write

$$f_0(\lambda) = 1, \quad f_1(\lambda) = a_1 - \lambda$$

and then use

$$f_r(\lambda) = (a_r - \lambda) f_{r-1}(\lambda) - b_{r-1} c_r f_{r-2}(\lambda)$$

for $r = 2, 3, \ldots, n$.

The corresponding formulas for the derivatives are:

$$f_0'(\lambda) = 0, \; f_1'(\lambda) = -1$$

$$f_r'(\lambda) = (a_r - \lambda) f_{r-1}'(\lambda) - b_{r-1} c_r f_{r-2}'(\lambda) - f_{r-1}(\lambda)$$

and these can be computed at the same time as the $f_r(\lambda)$. From Problem 5.13 (iv), or otherwise, the characteristic values of $\mathbf{A}$ are

$$0 + 20 \cos r\pi/6, \quad r = 1, 2, 3, 4, 5.$$

i.e.

$$\pm 10\sqrt{3}, \; \pm 10, \; 0.$$

The characteristic polynomial of $\mathbf{B}$ is

$$-\lambda^5 + * + 12\lambda^3 + 12\lambda^2 - 17\lambda - 18$$

$$= -(\lambda + 1)(\lambda + 2)(\lambda^3 - 3\lambda^2 - 5\lambda + 9)$$

and the characteristic values are

$$-2, \; -1.9459970, \; -1, \quad 1.2520004, \quad 3.6939948.$$

Since $\operatorname{tr} \mathbf{B} = 0$, the term in λ^4 being absent, some idea of the errors in our solution is given by the *actual* sum of the characteristic values which is 4×10^{-7}.

8.9. Solution

The successive reductions are

$$\begin{bmatrix} 5 & 9.2195446 & 0 & 5 \\ 9.2195446 & 17.9058830 & 1.2235291 & 11.1719188 \\ 0 & 1.2235291 & 2.0941177 & 2.2777696 \\ 5 & 11.1719188 & 2.2777696 & 10 \end{bmatrix}$$

$$\begin{bmatrix} 5 & 10.4880889 & 0 & 0 \\ 10.4880889 & 25.4727293 & 2.1614262 & 2.7806542 \\ 0 & 2.1614262 & 2.0941177 & 1.4189767 \\ 5 & 2.7806542 & 1.4189767 & 2.4331552 \end{bmatrix}$$

and that given. The rotations used are given by

$$c=.7592566 \quad s=.6507914$$

$$c=.8790491 \quad s=.4767313$$

$$c=.6137097 \quad s=.7895317.$$

As a check we compute the determinant of $\mathbf{W}_1$ using the recurrence method (p. 30) and we get

$$\det \mathbf{W}_1 = 1.0000053.$$

The characteristic roots of $\mathbf{W}$ are approximately

$$.0105, \quad .8431, \quad 3.858, \quad 30.29.$$

The Householder vectors are

$$[0, \ .9131, \ .3133, \ .2611]'$$

and

$$[0, 0, \ .8533, \ .5215]'.$$

8.11. Solution

See, e.g., H. Rutishauser, *On Jacobi rotation patterns,* pp. 219—239, in Proc. Symposia in Applied Mathematics, vol. 15 (American Mathematical Society, 1963).

8.12. Solution

The characteristic polynomial

$$\tilde{H}_6(\lambda) = \lambda^6 - 15\lambda^4 + 45\lambda^2 - 15$$

of this matrix is the one of the normalizations of the Hermite polynomial, $H_6(x)$. Specifically, it is clear that

$$\tilde{H}_n(\lambda) = -\lambda \tilde{H}_{n-1}(\lambda) - (n-1)\tilde{H}_{n-2}(\lambda).$$

Comparing this with the standard recurrence relation

$$H_n(x) = 2xH_{n-1}(x) - 2(n-1)H_{n-2}(x)$$

we see that

$$\tilde{H}_n(\lambda) = 2^{-n/2} H_n(-\lambda/\sqrt{2}).$$

The zeros of $H_6(x)$ are (National Bureau of Standards, *Handbook of Mathematical Functions*, 1964, p. 924)

$$\pm .43608, \pm 1.33585, \pm 2.35060$$

and those of $\tilde{H}_n(\lambda)$ are got by multiplying these by $\sqrt{2}$

$$\pm .61671, \pm 1.88918, \pm 3.32426.$$

8.13. Solution

The characteristic values are approximately

$$22.406872,\ 7.513724,\ 4.848950,\ 1.327045,\ -1.096595.$$

The dominant characteristic vector is

$$[\ .024588,\ .302396,\ .453215,\ .577177,\ .556385]'$$

and that corresponding to the characteristic value near 5 is

$$[-.547173,\ .312570,\ -.618112,\ .115607,\ .455494]'.$$

This example is due to J. H. Wilkinson (Numer. Math., *4,* 354—376 (1962)); see also John Todd (*Error in digital computations,* vol, 1, ed., L. B. Rall, 1965, pp. 3—41).

8.14. Solution

We want to have $\boldsymbol{A} = \boldsymbol{O}\boldsymbol{R}$, i.e. $\boldsymbol{O}'\boldsymbol{A} = \boldsymbol{R}$. It will be sufficient to show that $\omega = [x_1, \hat{\omega}]$ where $\omega'\omega = 1$ can be chosen so that

$$\begin{bmatrix} 1-2x_1^2 & -2x_1\hat{\omega}' \\ -2x_1\hat{\omega} & \hat{I}-2\hat{\omega}\hat{\omega}' \end{bmatrix} \begin{bmatrix} a_{11}, & \dots \\ \hat{\alpha}, & \dots \end{bmatrix} = \begin{bmatrix} b_{11}, & \dots \\ 0, & \dots \end{bmatrix} \tag{1}$$

for repetition of this process on successive principal submatrices will complete the triangulation of $\boldsymbol{A}$. (Note that this is an alternative to the Gaussian triangulation of $\boldsymbol{A}$.)

Since $\boldsymbol{I} - 2\omega\omega'$ is orthogonal we have

$$a_{11}^2 + \hat{\alpha}'\hat{\alpha} = b_{11}^2. \tag{2}$$

We can clearly assume $\hat{\alpha} \neq 0$; there is a choice of sign in b_{11}. Multiplying (1) out we find

$$(3) \qquad (1-2x_1^2)a_{11}-2x_1\hat{\omega}'\hat{\alpha}=b_{11}.$$

$$(4) \qquad -2a_{11}x_1\hat{\omega}+\hat{\alpha}-2\hat{\omega}\hat{\omega}'\hat{\alpha}=0.$$

Substituting in (4) from (3) we find

$$x_1\hat{\alpha}=(a_{11}+b_{11})\hat{\omega}$$

which gives

$$\hat{\omega}=(x_1/(a_{11}-b_{11}))\hat{\alpha}$$

so that $\hat{\omega}$ is determined when x_1 is, since b_{11} is given by (2). The fact that $x_1^2+\hat{\omega}'\hat{\omega}=1$ gives

$$x_1^2=(a_{11}-b_{11})^2/(\hat{\alpha}'\hat{\alpha}+(a_{11}-b_{11})^2)$$

and this determines x_1 up to a sign.

8.15. Solution

The characteristic values are approximately

12.109309, $-.535898$, $-.679823$, -1.000000, -2.429488, -7.464102.

The "even" values are exactly $-2(2-\sqrt{3})$, -1, $-2(2+\sqrt{3})$.

Chapter 9

9.1. Solution

(J.) The solution to this system is obviously $x=\begin{bmatrix}1\\1\end{bmatrix}$. We make any guess $\boldsymbol{x}^{(0)}$ and write $\boldsymbol{\varepsilon}_0^{(0)}=\boldsymbol{x}_0^{(0)}-\begin{bmatrix}1\\1\end{bmatrix}$. From the text, in the Jacobi case,

$$\boldsymbol{\varepsilon}^{(r+1)}=\begin{bmatrix}0&\frac{1}{2}\\\frac{1}{2}&0\end{bmatrix}\boldsymbol{\varepsilon}^{(r)},$$

$$\boldsymbol{\varepsilon}^{(r)}=2^{-r}\begin{bmatrix}0&1\\1&0\end{bmatrix}^r\boldsymbol{\varepsilon}^{(0)};$$

clearly

$$\begin{bmatrix}0&1\\1&0\end{bmatrix}^{2n}=\begin{bmatrix}1&0\\0&1\end{bmatrix},\quad \begin{bmatrix}0&1\\1&0\end{bmatrix}^{2n+1}=\begin{bmatrix}0&1\\1&0\end{bmatrix}$$

and we see that the components of $\boldsymbol{\varepsilon}^{(r)}$ are obtained from those of $\boldsymbol{\varepsilon}^{(0)}$ by dividing by 2^r and interchanging if r is odd.

(G.-S.) From the text, in this case,

$$\boldsymbol{\varepsilon}^{(r+1)}=\begin{bmatrix}1 & 0\\ -\frac{1}{2} & 1\end{bmatrix}^{-1}\begin{bmatrix}0 & \frac{1}{2}\\ 0 & 0\end{bmatrix}\boldsymbol{\varepsilon}^{(r)}$$

$$=\begin{bmatrix}1 & 0\\ \frac{1}{2} & 1\end{bmatrix}\begin{bmatrix}0 & \frac{1}{2}\\ 0 & 0\end{bmatrix}\boldsymbol{\varepsilon}^{(r)}$$

$$=\begin{bmatrix}0 & \frac{1}{2}\\ 0 & \frac{1}{4}\end{bmatrix}\boldsymbol{\varepsilon}^{(r)}.$$

It is clear that

$$\boldsymbol{\varepsilon}^{(r)}=\begin{bmatrix}2\cdot 2^{-2r}\,\varepsilon_2^{(0)}\\ 2^{-r}\,\varepsilon_2^{(0)}\end{bmatrix}.$$

9.2. Solution

The inverse is

$$\begin{bmatrix}1 & 0 & 0\\ a & 1 & 0\\ ac+b & c & 1\end{bmatrix}.$$

In the first case we have to find the spectral radii of

$$\begin{bmatrix}0 & 2 & -2\\ 1 & 0 & 1\\ 2 & 2 & 0\end{bmatrix}$$

and

$$\begin{bmatrix}1 & 0 & 0\\ -1 & 1 & 0\\ -2 & -2 & 1\end{bmatrix}^{-1}\begin{bmatrix}0 & 2 & -2\\ 0 & 0 & 1\\ 0 & 0 & 0\end{bmatrix}=\begin{bmatrix}1 & 0 & 0\\ 1 & 1 & 0\\ 4 & 2 & 1\end{bmatrix}\begin{bmatrix}0 & 2 & -2\\ 0 & 0 & 1\\ 0 & 0 & 0\end{bmatrix}=\begin{bmatrix}0 & 2 & -2\\ 0 & 2 & -1\\ 0 & 8 & -6\end{bmatrix}.$$

The first matrix has characteristic polynomial $-\lambda^3$ and the second $-\lambda\{\lambda^2+4\lambda-4\}$ so that the spectral radii are $0, 2(1+\sqrt{2})$ respectively. Thus, the Jacobi process converges and the Gauss—Seidel process does not.

In the second case we have to find the spectral radii of

$$\begin{bmatrix}0 & -\frac{1}{2} & \frac{1}{2}\\ 1 & 0 & 1\\ -\frac{1}{2} & \frac{1}{2} & 0\end{bmatrix}$$

and

$$\begin{bmatrix}1 & 0 & 0\\ -1 & 1 & 0\\ \frac{1}{2} & \frac{1}{2} & 1\end{bmatrix}^{-1}\begin{bmatrix}0 & -\frac{1}{2} & \frac{1}{2}\\ 0 & 0 & 1\\ 0 & 0 & 0\end{bmatrix}=\begin{bmatrix}1 & 0 & 0\\ 1 & 1 & 0\\ 0 & -\frac{1}{2} & 1\end{bmatrix}\begin{bmatrix}0 & -\frac{1}{2} & \frac{1}{2}\\ 0 & 0 & 1\\ 0 & 0 & 0\end{bmatrix}=\begin{bmatrix}0 & -\frac{1}{2} & \frac{1}{2}\\ 0 & -\frac{1}{2} & 3/2\\ 0 & 0 & -\frac{1}{2}\end{bmatrix}.$$

The first matrix has characteristic polynomial $-\lambda(\lambda^2+5/4)$ and the second $-\lambda(\lambda+\frac{1}{2})^2$ so that the spectral radii are $\sqrt{5}/2$ and $\frac{1}{2}$ respectively. Thus the Gauss—Seidel process converges and the Jacobi process does not.

9.3. Solution

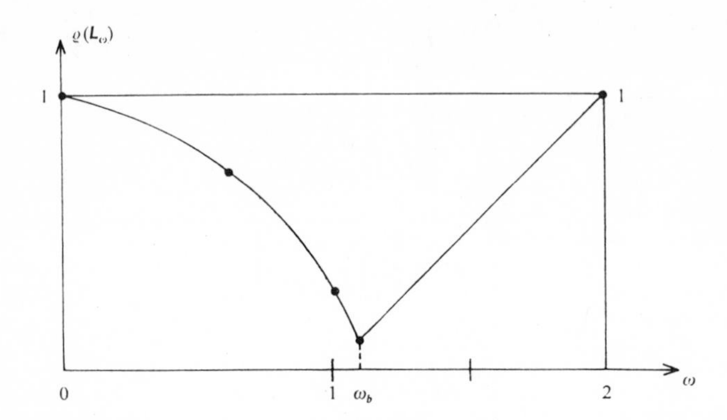

It is clear from the graph, and can be rigorously established, that the change in $\varrho(\mathcal{L}_\omega)$ in the neighborhood of the optimal ω is much larger for negative changes as for positive. This suggests that when we do not know the optimal ω, it is preferable to take an overestimate.

9.4. Solution

We find, for instance,

$$\mathbf{X}_1=\begin{bmatrix}-4.26 & 4.14 & -0.86\\ 4.14 & -5.06 & 1.94\\ -0.86 & 1.94 & -1.06\end{bmatrix},\quad \mathbf{X}_2=\begin{bmatrix}-3.9976 & 3.9864 & -1.0136\\ 3.9864 & -4.9896 & 2.0104\\ -1.0136 & 2.0104 & -0.9896\end{bmatrix},$$

so that

$$\mathbf{A}^{-1}-\mathbf{X}_1=\begin{bmatrix}0.26 & -0.14 & -0.14\\ -0.14 & 0.06 & 0.06\\ 0.14 & 0.06 & 0.06\end{bmatrix},$$

$$\mathbf{A}^{-1}-\mathbf{X}_2=\begin{bmatrix}0.0024 & 0.0136 & 0.0136\\ 0.0136 & -0.0104 & -0.0104\\ 0.0136 & -0.0104 & -0.0104\end{bmatrix}.$$

9.6. Solution

We want to find the roots of $\det(\mathbf{A}-\lambda\mathbf{I})=0$ where $\mathbf{A}=\begin{bmatrix}0 & \tau\mathbf{J}\\ \tau\mathbf{J} & 2\tau^2\mathbf{J}\end{bmatrix}$ and $\mathbf{J}=\begin{bmatrix}1 & 1\\ 1 & 1\end{bmatrix}$, $\tau=(1/4)\,\omega$, $\sigma=1-\omega$.

(a) From Problem 4.10 we have, since $\boldsymbol{J}^2=2\boldsymbol{J}$,

$$\det(\boldsymbol{A}-\lambda\boldsymbol{I})=\det(-2\lambda\tau^2\boldsymbol{J}+\lambda^2\boldsymbol{I}-2\sigma\tau^2\boldsymbol{J}),$$

i.e.

$$\det(\boldsymbol{A}-\lambda\boldsymbol{I})=\det\begin{bmatrix}-2\lambda\tau^2-2\sigma\tau^2+\lambda^2 & -2\lambda\tau^2-2\sigma\tau^2\\ -2\lambda\tau^2-2\sigma\tau^2 & -2\lambda\tau^2-2\sigma\tau^2+\lambda^2\end{bmatrix}$$

$$=\det\begin{bmatrix}\lambda^2 & -2\lambda\tau^2-2\sigma\tau^2\\ -\lambda^2 & -2\lambda\tau^2-2\sigma\tau^2+\lambda^2\end{bmatrix}\mathrm{col}_1-\mathrm{col}_2$$

$$=\lambda^2(\lambda^2-4\tau^2\lambda-4\sigma\tau^2).$$

Hence the characteristic roots are

$$0,\ 0,\ 2\tau^2\pm2\sqrt{\tau^4+\sigma\tau^2}.$$

(b) Alternatively, by Williamson's Theorem, Problem 6.11, the characteristic roots of $\boldsymbol{A}$ are those of

$$\begin{bmatrix}0 & 2\tau\\ 2\sigma\tau & 4\tau^2\end{bmatrix}\quad\text{and}\quad\begin{bmatrix}0 & 0\\ 0 & 0\end{bmatrix}$$

corresponding to the characteristic roots 2 and 0 of $\boldsymbol{J}$.

9.7. Solution

(J.) We may assume the matrix $\boldsymbol{A}$ normalized to have units on the diagonal. The condition for convergence of the Jacobi process is that $\varrho(\boldsymbol{I}-\boldsymbol{A})<1$. Since $\boldsymbol{A}$ is strictly diagonally dominant we have $\Lambda_i=\sum'|a_{ij}|<1$ for each i. This means that the Gerschgorin circles of $\boldsymbol{I}-\boldsymbol{A}$, which are all centered at the origin, have radii <1. Hence $\varrho(\boldsymbol{I}-\boldsymbol{A})<1$.

(G.—S.) We have now to show that $\varrho((\boldsymbol{I}-\boldsymbol{L})^{-1}\boldsymbol{U})<1$, again assuming normalization. If λ, x is a characteristic pair for $(\boldsymbol{I}-\boldsymbol{L})^{-1}\boldsymbol{U}$ then $(\boldsymbol{I}-\boldsymbol{L})^{-1}\boldsymbol{U}\boldsymbol{x}=$ $=\lambda\boldsymbol{x}$ which gives

$$(\boldsymbol{U}+\lambda\boldsymbol{L})\boldsymbol{x}=\lambda\boldsymbol{x}. \tag{1}$$

Let $x_M=\max|x_i|$. The M-th equation in (1) gives

$$\sum_{j<M}a_{Mj}x_j+\lambda\sum_{j>M}a_{Mj}x_j=\lambda x_M,$$

i.e.

$$\sum_{j<M}a_{Mj}(x_j/x_M)+\lambda\sum_{j>M}a_{Mj}(x_j/x_M)=\lambda. \tag{2}$$

If $|\lambda|\geqq1$ the relation (2) is impossible since the absolute value of the left hand side

$$\leqq \sum_{j<M} |a_{Mj}| + \lambda \sum_{j>M} |a_{Mj}|, \quad \text{by choice of } M,$$

$$\leqq \lambda \sum{}' |a_{Mj}|, \quad \text{since } \lambda \geqq 1,$$

$$< \lambda, \quad \text{since } \mathbf{A} \text{ is strictly diagonally dominant.}$$

Hence $|\lambda| < 1$, as required.

9.8. Solution

We begin by proving the following result:

If $\mathbf{F}$ and $\mathbf{G}$ are matrices such that $\mathbf{F}$ is non-singular and $\mathbf{F}+\mathbf{G}$ and $\mathbf{F}-\mathbf{G}^*$ are positive definite hermitian then the characteristic values of $\mathbf{F}^{-1}\mathbf{G}$ are all inside the unit circle.

Let λ, $\boldsymbol{x}$ be a characteristic pair for $\mathbf{F}^{-1}\mathbf{G}$. Then we find that $\boldsymbol{x}^*\mathbf{G}\boldsymbol{x} = \lambda \boldsymbol{x}^*\mathbf{F}\boldsymbol{x}$ and from this, by adding $\boldsymbol{x}^*\mathbf{F}\boldsymbol{x}$ to each side

$$\boldsymbol{x}^*(\mathbf{F}+\mathbf{G})\boldsymbol{x} = (1+\lambda)\boldsymbol{x}^*\mathbf{F}\boldsymbol{x}. \tag{1}$$

Since $\mathbf{F}+\mathbf{G}$ is positive definite hermitian, it follows that $\lambda \neq -1$.

Since $\mathbf{F}+\mathbf{G}$ is hermitian the right-hand side of (1) is equal to its conjugate transposed we have

$$(1+\bar{\lambda})\boldsymbol{x}^*\mathbf{F}^*\boldsymbol{x} = (1+\lambda)\boldsymbol{x}^*\mathbf{F}\boldsymbol{x}$$

$$= (1+\lambda)[\boldsymbol{x}^*(\mathbf{F}-\mathbf{G}^*)\boldsymbol{x} + \boldsymbol{x}^*\mathbf{G}^*\boldsymbol{x}]$$

$$= (1+\lambda)[\boldsymbol{x}^*(\mathbf{F}-\mathbf{G}^*)\boldsymbol{x} + \bar{\lambda}\boldsymbol{x}^*\mathbf{F}^*\boldsymbol{x}],$$

i.e.

$$(1-|\lambda|^2)\boldsymbol{x}^*\mathbf{F}^*\boldsymbol{x} = (1+\lambda)\boldsymbol{x}^*(\mathbf{F}-\mathbf{G}^*)\boldsymbol{x}.$$

Multiply across by $1+\bar{\lambda}$ and we get

$$(1-|\lambda|^2)\{(1+\bar{\lambda})\boldsymbol{x}^*\mathbf{F}\boldsymbol{x}\} = |1+\lambda|^2\boldsymbol{x}^*(\mathbf{F}-\mathbf{G}^*)\boldsymbol{x}.$$

The bracketed factors on the left can be replaced by $\boldsymbol{x}^*(\mathbf{F}+\mathbf{G})\boldsymbol{x}$ because $\mathbf{F}+\mathbf{G}$ is positive definite hermitian and we can star (1). Hence

$$(1-|\lambda|^2)\boldsymbol{x}^*(\mathbf{F}+\mathbf{G})\boldsymbol{x} = |1+\lambda|^2\boldsymbol{x}^*(\mathbf{F}-\mathbf{G}^*)\boldsymbol{x}. \tag{2}$$

The hermitian forms in (2) are positive if $\boldsymbol{x} \neq 0$. Hence $1-|\lambda|^2 > 0$ which is the result required.

The problem is now easily solved. Since $\mathbf{A}$ is hermitian $\mathbf{A} = \mathbf{L}+\mathbf{D}+\mathbf{L}^*$ with $\mathbf{D}$ real. If $\mathbf{F} = \mathbf{L}+\mathbf{D}$, $\mathbf{G} = \mathbf{L}^*$ then $\mathbf{A} = \mathbf{L}+\mathbf{G}$ is positive definite hermitian and so clearly is its diagonal $\mathbf{D} = \mathbf{F}-\mathbf{G}^*$. The result just established give the conclusion wanted.

9.9. Solution

$$\begin{bmatrix}0\\0\\0\\0\end{bmatrix};\ \begin{bmatrix}3.20\\.12\\.67\\.20\end{bmatrix};\ \begin{bmatrix}2.44\\.18\\1.06\\.35\end{bmatrix};\ \begin{bmatrix}1.97\\.21\\1.28\\.46\end{bmatrix};\ \begin{bmatrix}1.70\\.22\\1.39\\.55\end{bmatrix};\ \begin{matrix}1.55\\.21\\1.44\\.61\end{matrix};\ \ldots$$

9.10. Solution

p.	298	$.43\times10^{7}$	$.15\times10^{-1}$	$.58\times10^{1}$	$.99\times10^{-2}$	$.10\times10^{-1}$
p.	20	$.41\times10^{4}$	$.19\times10^{-4}$	$.57\times10^{-4}$	$.86\times10^{-5}$	$.86\times10^{-5}$
p.	21	$.16\times10^{8}$	$.13$	$.18\times10^{3}$	$.30\times10^{-1}$	$.30\times10^{-1}$
p.	300	$.13\times10^{10}$	$.18\times10^{-1}$	$.65\times10^{-1}$	$.96\times10^{-2}$	$.96\times10^{-2}$
p.	301	$.14\times10^{13}$	$.20\times10^{1}$	$.26\times10^{2}$	$.10\times10^{3}$	$.35\times10^{3}$
p.	301	$.16\times10^{16}$	$.39\times10^{2}$	$.16\times10^{4}$	$.59\times10^{3}$	$.85\times10^{3}$

(1) The condition number quoted is the one related to the euclidean vector norm; it would be more appropriate to take that associated with the Frobenius matrix norm.

(2) The $\|\boldsymbol{I}-\boldsymbol{A}\boldsymbol{X}\|_F$ are systematically smaller than $\|\boldsymbol{I}-\boldsymbol{X}\boldsymbol{A}\|_F$.

(3) When the inverses are reasonable (i.e. in the first four cases) refinement takes place; in the last two, there is a deterioration.

(4) Compare with the results of Problem 5.14.

Chapter 10

10.1. Solution

We shall show that the Rayleigh—Ritz process leads, in this case, to the same solution as the Galerkin process. We discuss the problem, initially, in a somewhat more general case than is necessary.

We begin by showing the relation of the variational problem

$$\min_{c\neq0}\frac{c'\boldsymbol{A}c}{c'\boldsymbol{B}c}$$

where $\boldsymbol{A}$ is symmetric and $\boldsymbol{B}$ is positive definite to the generalized characteristic value problem (of which a special case was handled in Problem 6.14):

$$\text{(1)}\qquad \boldsymbol{A}x=\lambda\boldsymbol{B}x.$$

We can write this in the form of a simple characteristic value problem

$$\mathbf{B}^{-1}\mathbf{A}\boldsymbol{x}=\lambda\boldsymbol{x}$$

but the matrix $\mathbf{B}^{-1}\mathbf{A}$ is no longer necessarily symmetric. We can, however, get a symmetric problem in the following way. From the LDU Theorem we can express $\mathbf{B}$ in the form $\mathbf{B}=\mathbf{T}\mathbf{T}'$ where $\mathbf{T}$ is a non-singular triangular matrix and then (1) can be written as

$$((\mathbf{T}^{-1})'\mathbf{A}\mathbf{T}^{-1}-\lambda\mathbf{I})\boldsymbol{y}=0, \quad \text{where} \quad \boldsymbol{y}=\mathbf{T}\boldsymbol{x}.$$

Methods for the solution of the symmetric case apply but we have the additional computation of $\boldsymbol{x}=\mathbf{T}^{-1}\boldsymbol{y}$. Note that the characteristic vectors $\boldsymbol{y}$ are no longer orthogonal in the ordinary sense, but that we have $\boldsymbol{x}_1'\mathbf{B}\boldsymbol{x}_2=0$ if $\boldsymbol{x}_1, \boldsymbol{x}_2$ correspond to distinct λ_1, λ_2. We can now essentially repeat the argument of Problem 1.9. We expand any $\boldsymbol{c}$ in the form $\boldsymbol{c}=\sum \gamma_i \boldsymbol{x}_i$ and then note that

$$\frac{\boldsymbol{c}'\mathbf{A}\boldsymbol{c}}{\boldsymbol{c}'\mathbf{B}\boldsymbol{c}}=\frac{\sum \lambda_i \gamma_i^2}{\sum \gamma_i^2}$$

which shows that

$$\lambda_1 \leqq R(\boldsymbol{c}) \leqq \lambda_n$$

if λ_1 and λ_n are the extreme generalized characteristic values.

We note, as in the simple case, that $R(\boldsymbol{c})$ has an extreme point for $\boldsymbol{c}=\boldsymbol{x}_i$, with value λ_i. This can be proved by the use of Lagrange multipliers.

We now return to the problem proper. In the case of the system (1) when we use the basis functions $b_r(x)$ suggested we are led to the problem

$$\min_{\boldsymbol{\beta}\neq 0} \frac{\boldsymbol{\beta}'\mathbf{A}\boldsymbol{\beta}}{\boldsymbol{\beta}'\mathbf{B}\boldsymbol{\beta}}$$

where $\mathbf{A}, \mathbf{B}$ are the same triple diagonal matrices. In view of what has just been proved we come again to the same generalized characteristic value problem and to the same results as before.

Several questions are apparent and form the beginning of a deeper study of the Rayleigh—Ritz method:

What are good ways to choose the b's?

How does the convergence to the characteristic values and vectors depend on n?

Can lower bounds on the characteristic values be obtained?

Chapter 11

11.1. Solution

(1) In the notation of the text we have

$$\boldsymbol{f}=\begin{bmatrix}1\\3\\1\end{bmatrix},\quad \boldsymbol{Q}=\begin{bmatrix}1&1\\1&2\\1&3\end{bmatrix}.$$

We find

$$\boldsymbol{Q}'\boldsymbol{Q}=\begin{bmatrix}3&6\\6&14\end{bmatrix},\quad (\boldsymbol{Q}'\boldsymbol{Q})^{-1}=\frac{1}{6}\begin{bmatrix}14&-6\\-6&3\end{bmatrix},$$

$$\boldsymbol{Q}'\boldsymbol{f}=\begin{bmatrix}5\\10\end{bmatrix},\quad c\doteq\frac{1}{6}\begin{bmatrix}14&-6\\-6&3\end{bmatrix}\begin{bmatrix}5\\10\end{bmatrix}=\begin{bmatrix}5/3\\0\end{bmatrix},$$

so that $y=5/3$ is the best approximation.

(2) Alternatively we take $\boldsymbol{Q}'\boldsymbol{Q}=\boldsymbol{L}\boldsymbol{L}'$ where

$$\boldsymbol{L}=\begin{bmatrix}\sqrt{3}&0\\2\sqrt{3}&\sqrt{2}\end{bmatrix}$$ and we solve first $\boldsymbol{L}\,y=\begin{bmatrix}5\\10\end{bmatrix}$ getting $y=\begin{bmatrix}5/\sqrt{3}\\0\end{bmatrix}$

and then $\boldsymbol{L}'c=y$ getting $c=\begin{bmatrix}5/3\\0\end{bmatrix}$ as before.

(3) An analytic solution to the problem is as follows: Assume $y=ax+b$ to be the linear fit. Then

$$E=(a+b-1)^2+(2a+b-3)^2+(3a+b-1)^2$$

which gives

$$\frac{\partial E}{\partial a}=28a+12b-20,$$

$$\frac{\partial E}{\partial b}=12a+6b-10.$$

Solving

$$7a+3b=5 \quad\text{and}\quad 6a+3b=5$$

we find $a=0$, $b=5/3$, as before.

(4) Consider the over-determined system

$$\left.\begin{aligned}x+\ \,y&=1\\x+2y&=3\\x+3y&=1\end{aligned}\right\}.$$

The factorization $\mathbf{Q}=\boldsymbol{\Phi}\mathbf{U}$ is:

$$\begin{bmatrix}1, & 1\\ 1, & 2\\ 1, & 3\end{bmatrix}=\begin{bmatrix}1/\sqrt{3}, & -1/\sqrt{2}\\ 1/\sqrt{3}, & 0\\ 1/\sqrt{3}, & 1/\sqrt{2}\end{bmatrix}\begin{bmatrix}\sqrt{3}, & 2\sqrt{3}\\ 0, & \sqrt{2}\end{bmatrix}$$

and we find first

$$\boldsymbol{y}=\begin{bmatrix}1/\sqrt{3}, & 1/\sqrt{3}, & 1/\sqrt{3}\\ -1/\sqrt{2}, & 0, & 1/\sqrt{2}\end{bmatrix}\begin{bmatrix}1\\ 3\\ 1\end{bmatrix}=\begin{bmatrix}5/\sqrt{3}\\ 0\end{bmatrix}$$

and then solve $\mathbf{U}\boldsymbol{c}=\boldsymbol{y}$, i.e.,

$$\begin{bmatrix}\sqrt{3}, & 2\sqrt{3}\\ 0, & \sqrt{2}\end{bmatrix}\boldsymbol{c}=\begin{bmatrix}5/\sqrt{3}\\ 0\end{bmatrix}$$

giving $\boldsymbol{y}'=[5/3, 0]$, as before.

11.2, 11.3. Solution

In connection with these problems see papers by R. H. Wampler, in particular *An evaluation of linear least squares computer programs,* J. Research National Bureau Stand. *73B*, 59—90 (1969).

This paper gives some idea of the variable quality of library subroutines.

11.4. Solution

It is geometrically obvious and easily proved that the line in question is such that the residuals r_1, r_2, r_3 where

$$r_i=ax_i+b-f_i$$

are equal in magnitude but alternate in sign. Thus the best fit is given by

$$y=2.$$

For methods of handling this problem when there are more than three points x_i see F. Scheid, *The under-over-under theorem,* Amer. Math. Monthly, *68*, 862—871 (1961).

11.5. Solution

$$E=(x+y-1)^2+(2x+2y-0)^2+(-x-y-2)^2.$$

$$\frac{1}{2}\cdot\frac{\partial E}{\partial x}=6(x+y)+1;\quad \frac{1}{2}\cdot\frac{\partial E}{\partial y}=6(x+y)+1.$$

Hence $x+y=-1/6$ for a minimum which is 170/36.

11.6. Solution

$$E=(x+y-1)^2+(2x-0)^2+(-x+3y-2)^2.$$

$$\frac{1}{2}\cdot\frac{\partial E}{\partial x}=6x-2y-1;\quad \frac{1}{2}\cdot\frac{\partial E}{\partial y}=-2x+10y-7.$$

Hence $x=1/14$, $y=5/7$ for a minimum which is $1/14$.

Chapter 12

12.1. Solution

We may assume that the leading $r\times r$ submatrix $\mathbf{A}_{11}$ of $\mathbf{A}$ is non-singular. [For if not there are permutation matrices $\mathbf{P}, \mathbf{Q}$ such that $\mathbf{PAQ}$ has this property and if $\mathbf{PAQ}=\mathbf{BC}$ then $\mathbf{A}=(\mathbf{P}^{-1}\mathbf{B})(\mathbf{C}\mathbf{Q}^{-1})$.]

We observe that

$$\mathbf{A}=\begin{bmatrix}\mathbf{A}_{11} & \mathbf{A}_{12}\\ \mathbf{A}_{21} & \mathbf{A}_{22}\end{bmatrix}=\begin{bmatrix}\mathbf{I}\\ \mathbf{A}_{21}\mathbf{A}_{11}^{-1}\end{bmatrix}[\mathbf{A}_{11}\ \ \mathbf{A}_{12}].$$

We only have to justify the equality $\mathbf{A}_{21}\mathbf{A}_{11}^{-1}\mathbf{A}_{12}=\mathbf{A}_{22}$. Since $\mathbf{A}$ has rank r and since its first r rows are independent the remaining $m-r$ rows can be represented as linear combinations of them. That is, there is an $m-r\times r$ matrix $\mathbf{X}$ such that

$$\mathbf{X}[\mathbf{A}_{11}\ \ \mathbf{A}_{12}]=[\mathbf{A}_{21}\ \ \mathbf{A}_{22}],$$

i.e. $\mathbf{X}\mathbf{A}_{11}=\mathbf{A}_{21}$ and $\mathbf{X}\mathbf{A}_{12}=\mathbf{A}_{22}$; the required equality follows. Finally, note that the rank of each of the two factors of $\mathbf{A}$ is exactly r: the first which has r columns includes $\mathbf{I}_r$ and the second, which has r rows, includes $\mathbf{A}_{11}$.

Clearly, if $\mathbf{B}, \mathbf{C}$ are possible factors, so are $\mathbf{BM}$ and $\mathbf{M}^{-1}\mathbf{C}$ for any non-singular $r\times r$ matrix $\mathbf{M}$. On the other hand, if

$$\mathbf{A}=\mathbf{BC}=\mathcal{B}\mathcal{C}$$

then

$$\mathbf{B}'\mathbf{BC}=B'\mathcal{B}\mathcal{C}$$

and so $\mathbf{C}=[(\mathbf{B}'\mathbf{B})^{-1}\mathbf{B}'\mathcal{B}]\mathcal{C}=\mathbf{M}\mathcal{C}$, say. The matrix $\mathbf{B}'\mathbf{B}$ is non-singular since it has rank r with $\mathbf{B}$. Since $\mathbf{C}, \mathcal{C}$ each have rank r, so has $\mathbf{M}$ and we can therefore write $\mathcal{C}=\mathbf{M}^{-1}\mathbf{C}$ as required. Continuing, from $\mathbf{C}=\mathbf{M}\mathcal{C}$ follows

$$\mathcal{B}\mathcal{C}=\mathbf{BC}=\mathbf{BM}\mathcal{C}$$

and so $\mathcal{B}\mathcal{C}\mathcal{C}'=\mathbf{BM}\mathcal{C}\mathcal{C}'$ which gives $\mathcal{B}=\mathbf{BM}$ as required since we can post-multiply across by $(\mathcal{C}\mathcal{C}')^{-1}$ for $\mathcal{C}\mathcal{C}'$ is of full rank with $\mathcal{C}$.

REMARK. We have here used two standard results about rank:

(1) rank $\mathbf{AB} \leq \min [\text{rank } \mathbf{A}, \text{rank } \mathbf{B}]$,

(2) rank $\mathbf{AA}^* = \text{rank } \mathbf{A}^*\mathbf{A} = \text{rank } \mathbf{A} = \text{rank } \mathbf{A}^*$ for matrices with complex elements.

We include a proof of (2).

Clearly $\mathbf{A}^*\mathbf{A}\mathbf{x}=0$ implies $\mathbf{x}^*\mathbf{A}^*\mathbf{A}\mathbf{x}=0$ and so $\mathbf{A}\mathbf{x}=0$. On the other hand $\mathbf{A}\mathbf{x}=0$ implies $\mathbf{A}^*\mathbf{A}\mathbf{x}=0$. Thus the null spaces $\mathcal{N}(\mathbf{A})$, $\mathcal{N}(\mathbf{A}^*\mathbf{A})$ are the same. Now it is a standard result that

(3) $\dim \mathcal{N}(\mathbf{M}) = \text{number of columns of } \mathbf{M} - \text{rank } \mathbf{M}$.

Now $\mathbf{A}^*\mathbf{A}$ and $\mathbf{A}$ have the same number of columns and so, necessarily, the same rank.

We outline a proof of (3).

Suppose an $n \times n$ matrix $\mathbf{A}$ has rank r. Without loss of generality we may assume that the leading $r \times r$ submatrix $\mathbf{A}_{11}$ is non-singular.

$$\mathbf{A} = \begin{bmatrix} \mathbf{A}_{11} & \mathbf{A}_{12} \\ \mathbf{A}_{21} & \mathbf{A}_{22} \end{bmatrix}.$$

Consider the $n-r$ vectors in V_n:

$$\boldsymbol{\zeta}_j = \begin{bmatrix} -\mathbf{A}_{11}^{-1}\mathbf{A}_{12}\mathbf{e}_j \\ \mathbf{f}_j \end{bmatrix}, \quad j=1, 2, \ldots, n-r,$$

where $\mathbf{e}_j$ is the j-th unit-vector in V_{n-r}. These matrices are clearly linearly independent and hence $\mathcal{M} = \text{span } (\zeta_1, \ldots, \zeta_{n-r})$ has dimension $n-r$ in V_n. We prove that the null space $\mathcal{N}(\mathbf{A})$ of $\mathbf{A}$ is exactly $\mathcal{M}$.

Since

$$\mathbf{A}\boldsymbol{\zeta}_j = \begin{bmatrix} \mathbf{A}_{11}(-\mathbf{A}_{11}^{-1}\mathbf{A}_{12}\mathbf{e}_j) + \mathbf{A}_{11}\mathbf{e}_j \\ \text{linear combinations of above } r \text{ rows} \end{bmatrix} = \begin{bmatrix} 0 \\ 0 \end{bmatrix}$$

it follows that each $\boldsymbol{\zeta}_j \in \mathcal{N}(\mathbf{A})$ and so $\mathcal{M} \subseteq \mathcal{N}(\mathbf{A})$ as $\mathcal{N}$ is a subspace.

To prove the opposite inclusion $\mathcal{M} \supseteq \mathcal{N}(\mathbf{A})$ take $\mathbf{x} = [\mathbf{x}_1, \mathbf{x}_2]' \in \mathcal{N}(\mathbf{A})$. Let $\mathbf{x}_2 = \sum c_j \mathbf{e}_j$. Then $\mathbf{A}\mathbf{x}=0$ gives $\mathbf{A}_{11}\mathbf{x}_1 + \mathbf{A}_{12}\mathbf{x}_2 = 0$, i.e. $\mathbf{x}_1 = -\mathbf{A}_{11}^{-1}\mathbf{A}_{12}\mathbf{x}_2$ and so, by definition of $\boldsymbol{\zeta}_j$, we have $\mathbf{x} = \sum c_j \boldsymbol{\zeta}_j$, i.e. $\mathbf{x} \in \mathcal{N}$.

12.2. Solution

(Problem 11.6.) The matrix $\mathbf{A} = \begin{bmatrix} 1 & 1 \\ 2 & 2 \\ -1 & -1 \end{bmatrix}$ has rank 1 and can be factorized as

$$\mathbf{A} = \begin{bmatrix} 1 \\ 2 \\ -1 \end{bmatrix} [1, \ 1] = \mathbf{BC}.$$

Here $\boldsymbol{B}'\boldsymbol{B}=[6]$, $\boldsymbol{C}\boldsymbol{C}'=[2]$ and

$$x=\begin{bmatrix}1\\1\end{bmatrix}\begin{bmatrix}\frac{1}{2}\end{bmatrix}\begin{bmatrix}\frac{1}{6}\end{bmatrix}[1,\ 2,\ -1]\begin{bmatrix}1\\0\\2\end{bmatrix}=-\begin{bmatrix}1/12\\1/12\end{bmatrix}.$$

Observe that $x_1+x_2=-1/6$ and that the solution obtained is that which has minimum length.

(Problem 11.6, second solution.)

We find, if $\boldsymbol{A}^*=\begin{bmatrix}1 & 2 & -1\\1 & 2 & -1\end{bmatrix}$ that

$$\boldsymbol{A}\boldsymbol{A}^*=\begin{bmatrix}2 & 4 & -2\\4 & 8 & -4\\-2 & -4 & 2\end{bmatrix}\quad\text{and}\quad \boldsymbol{A}^*\boldsymbol{A}=\begin{bmatrix}6 & 6\\6 & 6\end{bmatrix}.$$

We see that the characteristic values of $\boldsymbol{A}\boldsymbol{A}^*$ are 12, 0, 0 and those of $\boldsymbol{A}^*\boldsymbol{A}$ are 12, 0 (cf. Problem 6.12). The orthogonal similarities which diagonalize these matrices are easily found to be

$$\boldsymbol{U}^*(\boldsymbol{A}\boldsymbol{A}^*)\boldsymbol{U}=\operatorname{diag}\ [12,\ 0,\ 0]\quad\text{where}\quad \boldsymbol{U}=\begin{bmatrix}1/\sqrt{6} & -1/\sqrt{3} & 1/\sqrt{2}\\2/\sqrt{6} & 1/\sqrt{3} & 0\\-1/\sqrt{6} & 1/\sqrt{3} & 1/\sqrt{2}\end{bmatrix}$$

and

$$\boldsymbol{V}^*(\boldsymbol{A}^*\boldsymbol{A})\boldsymbol{V}=\operatorname{diag}\ [12,\ 0]\quad\text{where}\quad \boldsymbol{V}=\begin{bmatrix}1/\sqrt{2} & 1/\sqrt{2}\\1/\sqrt{2} & -1/\sqrt{2}\end{bmatrix}.$$

Thus the singular value decomposition of $\boldsymbol{A}$ is

$$\boldsymbol{A}=\boldsymbol{U}\Sigma\boldsymbol{V}^*,\quad \Sigma=\begin{bmatrix}\sqrt{12} & 0\\0 & 0\\0 & 0\end{bmatrix},$$

and the pseudo-inverse of $\boldsymbol{A}$ is

$$\boldsymbol{A}^I=\boldsymbol{V}^*\Sigma^I\boldsymbol{U}^*,\quad \Sigma^I=\begin{bmatrix}1/\sqrt{12} & 0 & 0\\0 & 0 & 0\end{bmatrix},$$

i.e.

$$\boldsymbol{A}^I=\frac{1}{12}\begin{bmatrix}1 & 2 & -1\\1 & 2 & -1\end{bmatrix}\quad\text{giving}\quad x=\boldsymbol{A}^I\begin{bmatrix}1\\0\\2\end{bmatrix}=\frac{1}{12}\begin{bmatrix}-1\\-1\end{bmatrix}\quad\text{as before.}$$

(Problem 11.7.) The matrix $\begin{bmatrix}1 & 1\\2 & 0\\-1 & 3\end{bmatrix}$ has rank 2 and we may take $\boldsymbol{B}=\boldsymbol{A}$,

$\mathbf{C}=\mathbf{I}$. Then

$$\begin{aligned}\mathbf{X}&=(\mathbf{A}'\mathbf{A})^{-1}\mathbf{A}'\mathbf{b}\\&=\begin{bmatrix}6&-2\\-2&10\end{bmatrix}^{-1}\begin{bmatrix}-1\\7\end{bmatrix}\\&=\frac{1}{56}\begin{bmatrix}10&2\\2&6\end{bmatrix}\begin{bmatrix}-1\\7\end{bmatrix}\\&=\frac{1}{14}\begin{bmatrix}1\\10\end{bmatrix}.\end{aligned}$$

12.3. Solution

$$225\,\mathbf{A}\mathbf{A}^*=\begin{bmatrix}425&-250&350\\-250&200&-100\\350&-100&500\end{bmatrix}=25\begin{bmatrix}17&-10&14\\-10&8&-4\\14&-4&20\end{bmatrix}.$$

The characteristic polynomial of $9\mathbf{A}\mathbf{A}^*$ is

$$\det\begin{bmatrix}17-\lambda&-10&14\\-10&8-\lambda&-4\\14&-4&20-\lambda\end{bmatrix}=\lambda(\lambda-9)(\lambda-36).$$

Hence the singular values of $\mathbf{A}$ are 0, 1, 2. The singular value decomposition is

$$\mathbf{A}=\mathbf{V}\begin{bmatrix}1&0\\0&2\\0&0\end{bmatrix}\mathbf{U}^*$$

where

$$\mathbf{V}=\frac{1}{3}\begin{bmatrix}1&-2&2\\-2&1&2\\-2&-2&-1\end{bmatrix},\quad \mathbf{U}^*=\frac{1}{5}\begin{bmatrix}3&-4\\4&3\end{bmatrix}.$$

Here $\mathbf{V}$ is the orthogonal matrix which diagonalizes $\mathbf{A}\mathbf{A}^*$, $\mathbf{V}^*\mathbf{A}\mathbf{A}^*\mathbf{V}=\text{diag.}[1,\ 2,\ 0]$ and $\mathbf{U}$ is the orthogonal matrix which diagonalizes $\mathbf{A}^*\mathbf{A}$:

$$\mathbf{A}^*\mathbf{A}=\frac{1}{25}\begin{bmatrix}73&36\\36&52\end{bmatrix}$$

$$\frac{1}{5}\begin{bmatrix}3&-4\\4&3\end{bmatrix}(\mathbf{A}^*\mathbf{A})\frac{1}{5}\begin{bmatrix}3&4\\-4&3\end{bmatrix}=\begin{bmatrix}1&0\\0&4\end{bmatrix}.$$

Hence

$$\begin{aligned}\mathbf{A}^I&=\frac{1}{15}\begin{bmatrix}3&4\\-4&3\end{bmatrix}\begin{bmatrix}1&0&0\\0&\frac{1}{2}&0\end{bmatrix}\begin{bmatrix}1&-2&-2\\-2&1&-2\\2&2&-1\end{bmatrix}\\&=\frac{1}{15}\begin{bmatrix}-1&4&-10\\-7&19/2&5\end{bmatrix}.\end{aligned}$$

In a similar way we find

$$\boldsymbol{B}^I=\frac{1}{11}\begin{bmatrix}2 & 6 & 9\\ 6 & 7 & -6\\ 9 & -6 & 2\end{bmatrix}\times\begin{bmatrix}1 & 0\\ 0 & 2\\ 0 & 0\end{bmatrix}\times\frac{1}{5}\begin{bmatrix}3 & -4\\ 4 & 3\end{bmatrix}$$

so that

$$\boldsymbol{B}^I=\frac{1}{55}\begin{bmatrix}18 & 32 & 15\\ 1 & -27/2 & -45\end{bmatrix}.$$

12.4. Solution

(1) Clearly

$$\boldsymbol{A}\boldsymbol{A}^*=\begin{bmatrix}3 & -1 & 2\\ -1 & 2 & 1\\ 2 & 1 & 3\end{bmatrix}.$$

$\boldsymbol{A}\boldsymbol{A}^*$ has characteristic values 5, 3, 0 and if

$$\boldsymbol{U}^*=\begin{bmatrix}1/\sqrt{2} & 0 & 1/\sqrt{2}\\ 1/\sqrt{6} & -2/\sqrt{6} & -1/\sqrt{6}\\ 1/\sqrt{3} & 1/\sqrt{3} & -1/\sqrt{3}\end{bmatrix}$$

then

$$\boldsymbol{U}^*\boldsymbol{A}\boldsymbol{A}^*\boldsymbol{U}=\begin{bmatrix}5 & 0 & 0\\ 0 & 3 & 0\\ 0 & 0 & 0\end{bmatrix}.$$

Hence

$$\boldsymbol{F}=\begin{bmatrix}\sqrt{2} & 1/\sqrt{2} & 1/\sqrt{2} & \sqrt{2}\\ 0 & -3/\sqrt{6} & 3/\sqrt{6} & 0\\ 0 & 0 & 0 & 0\end{bmatrix}$$

and

$$\boldsymbol{V}_1^*=\begin{bmatrix}\sqrt{2/5} & 1/\sqrt{10} & 1/\sqrt{10} & \sqrt{2/5}\\ 0 & -1/\sqrt{2} & 1/\sqrt{2} & 0\end{bmatrix}.$$

We then observe that we may take

$$\boldsymbol{V}_2^*=\begin{bmatrix}1/\sqrt{2} & 0 & 0 & -1/\sqrt{2}\\ -1/\sqrt{10} & 2/\sqrt{10} & 2/\sqrt{10} & -1/\sqrt{10}\end{bmatrix}$$

and then

$$\Sigma=\begin{bmatrix}\sqrt{2} & 1/\sqrt{2} & 1/\sqrt{2} & \sqrt{2}\\ 0 & -3/\sqrt{6} & 3/\sqrt{6} & 0\\ 0 & 0 & 0 & 0\end{bmatrix}\begin{bmatrix}\sqrt{2/5} & 0 & 1/\sqrt{2} & -1/\sqrt{10}\\ 1/\sqrt{10} & -1/\sqrt{2} & 0 & 2/\sqrt{10}\\ 1/\sqrt{10} & 1/\sqrt{2} & 0 & 2/\sqrt{10}\\ \sqrt{2/5} & 0 & -1/\sqrt{2} & -1/\sqrt{10}\end{bmatrix}=$$

$$=\begin{bmatrix}\sqrt{5} & 0 & 0 & 0\\ 0 & \sqrt{3} & 0 & 0\\ 0 & 0 & 0 & 0\end{bmatrix}.$$

(2) $\boldsymbol{A}$ has obviously rank 2 and can be factorized in the form

$$\boldsymbol{A}=\begin{bmatrix}1&0\\0&1\\1&1\end{bmatrix}\begin{bmatrix}1&0&1&1\\0&1&-1&0\end{bmatrix},$$

where each factor, again, obviously has rank 2. (Cf. Problem 12.1.)

Using the formulas of the text we find:

$$\boldsymbol{A}^I=\begin{bmatrix}1&0\\0&1\\1&-1\\1&0\end{bmatrix}\left[\begin{bmatrix}1&0&1&1\\0&1&-1&0\end{bmatrix}\begin{bmatrix}1&0\\0&1\\1&-1\\1&0\end{bmatrix}\right]^{-1}\left[\begin{bmatrix}1&0&1\\0&1&1\end{bmatrix}\begin{bmatrix}1&0\\0&1\\1&1\end{bmatrix}\right]^{-1}\begin{bmatrix}1&0&1\\0&1&1\end{bmatrix}$$

$$=\begin{bmatrix}1&0\\0&1\\1&-1\\1&0\end{bmatrix}\begin{bmatrix}3&-1\\-1&2\end{bmatrix}^{-1}\begin{bmatrix}2&1\\1&2\end{bmatrix}^{-1}\begin{bmatrix}1&0&1\\0&1&1\end{bmatrix}$$

$$=\frac{1}{15}\begin{bmatrix}1&0\\0&1\\1&-1\\1&0\end{bmatrix}\begin{bmatrix}2&1\\1&3\end{bmatrix}\begin{bmatrix}2&-1\\-1&2\end{bmatrix}\begin{bmatrix}1&0&1\\0&1&1\end{bmatrix}$$

$$=\frac{1}{15}\begin{bmatrix}2&1\\1&3\\1&-2\\2&1\end{bmatrix}\begin{bmatrix}2&-1&1\\-1&2&1\end{bmatrix}=\frac{1}{15}\begin{bmatrix}3&0&3\\-1&5&4\\4&-5&-1\\3&0&3\end{bmatrix}.$$

[This matrix was discussed by M. R. Hestenes, J. SIAM, *6*, 51—90 (1958).]

12.5. Solution

(a) Trivial. We have $\boldsymbol{A}=\boldsymbol{U}\boldsymbol{\Sigma}\boldsymbol{V}^*=\boldsymbol{U}\boldsymbol{V}^*\boldsymbol{V}\boldsymbol{\Sigma}\boldsymbol{V}^*=(\boldsymbol{U}\boldsymbol{V}^*)(\boldsymbol{V}\boldsymbol{\Sigma}\boldsymbol{V}^*)$ and $\boldsymbol{U}\boldsymbol{V}^*$ is unitary and $\boldsymbol{V}\boldsymbol{\Sigma}\boldsymbol{V}^*$ is positive definite.

(b) We have to show that

$$\text{(1)}\qquad \|\boldsymbol{A}-\boldsymbol{U}\boldsymbol{V}^*\|_F\leqq\|\boldsymbol{A}-\boldsymbol{W}\|_F \quad \text{for any unitary } \boldsymbol{W}.$$

Since $\|\boldsymbol{X}\|_F=\|\boldsymbol{U}_1\boldsymbol{X}\boldsymbol{U}_2\|_F$ for any unitary $\boldsymbol{U}_1$, $\boldsymbol{U}_2$ (Problem 2.10) (1) is equivalent to

$$\|\boldsymbol{\Sigma}-\boldsymbol{I}\|_F\leqq\|\boldsymbol{\Sigma}-\boldsymbol{U}^*\boldsymbol{W}\boldsymbol{V}\|_F \quad \text{for any unitary } \boldsymbol{W},$$

or to

$$\|\boldsymbol{A}-\boldsymbol{I}\|_F\leqq\|\boldsymbol{\Sigma}-\boldsymbol{W}_1\|_F \quad \text{for any unitary } \boldsymbol{W}_1.$$

Now

$$\|\Sigma - W_1\|_F^2 = \text{tr}\,(\Sigma - W_1)(\Sigma - W_1^*)$$
$$= \text{tr}\,(\Sigma^2 - \Sigma W_1^* - W_1 \Sigma + I)$$
$$= \Sigma \left(\sigma_r^2 - \sigma_r (\overline{\omega}_r + \omega_r) + 1\right)$$

where $\omega_1, \ldots, \omega_n$ are the diagonal elements of W_1. Now, W_1 being unitary, $|\omega_r| \leqq 1$ and so $(\omega_r^* + \omega_r) = 2\,\text{Re}\,\omega_r$ lies between ± 2 and

$$\|\Sigma - W_1\|_F^2 \geqq \Sigma (\sigma_r^2 - 2\sigma_r + 1) = \Sigma (\sigma_r - 1)^2 = \|\Sigma - I\|_F^2.$$

12.6. Solution

We outline, from first principles, the construction of A^I when $A \neq 0$ is a column vector a. Then AA^* is an $n \times n$ matrix with characteristic values $a^* a, 0, 0, \ldots, 0$ (cf. Problem 6.12). Since $(a a^*)\, a = a (a^* a)$, the characteristic vector of AA^* corresponding to $a^* a$ is a. This means that the first column of the unitary matrix U is $a/\sqrt{a^* a}$ — we shall see that no further information about U is required. Since $A^* A$ is a 1×1 matrix $[a^* a]$ we can take $V = [1]$. It is also clear that

$$\Sigma = \left[\sqrt{a^* a}, 0, 0, \ldots, 0\right]^*, \quad \Sigma^I = \left[1/\sqrt{a^* a}, 0, 0, \ldots, 0\right].$$

We have

$$A = U \Sigma V^* = \left[a/\sqrt{a^* a}, \ldots\right] \begin{bmatrix} \sqrt{a^* a} \\ 0 \\ 0 \\ \vdots \\ 0 \end{bmatrix} [1]$$

so that

$$A^I = V \Sigma^I U^* = [1] \left[1/\sqrt{a^* a}, 0, 0, \ldots, 0\right] \begin{bmatrix} a^*/\sqrt{a^* a} \\ \vdots \end{bmatrix}$$
$$= \frac{1}{a^* a} \cdot a^*.$$

It is easy to verify that all the axioms (12.4—12.7) are satisfied.

The pseudo-inverse of a zero vector is itself.

The results of this problem can also be found by applying (12.18) and taking $B = A$, $C = [1]$.

Bibliographical Remarks

Recommended Literature

T. M. Apostol, *Calculus*, I, II (Wiley, 1967—9).

A. Ostrowski, *Vorlesungen über Differential- und Integralrechnung*, 3 vols. *Aufgaben sammlung zur Infinitesimalrechnung*, 3 vols. (Birkhauser, 1965—72).

Gilbert Strang, *Linear algebra and its applications* (Academic Press, 1976).

Texts and Monographs

E. K. Blum, *Numerical analysis and computation, theory and practice* (Prentice Hall, 1972).

E. Bodewig, *Matrix calculus* (North Holland, 1956).

S. D. Conte and C. de Boor, *Elementary numerical analysis* (McGraw-Hill, 1973).

D. K. Faddeev and V. N. Faddeeva, tr. R. C. Williams, *Computational methods of linear algebra* (Freeman, 1963).

D. K. Faddeev and I. S. Sominskii, tr. J. L. Brenner, *Problems in higher algebra* (Freeman, 1965).

V. N. Faddeeva, tr. C. D. Benster, *Computational methods of linear algebra* (Dover, 1959).

G. E. Forsythe and C. B. Moler, *Computer solution of linear algebraic systems* (Prentice Hall, 1967).

L. Fox, *Introduction to numerical linear algebra* (Clarendon Press, 1966).

F. R. Gantmacher, tr. K. A. Hirsch, *Matrix theory*, 2 vols. (Chelsea, 1959).

N. Gastinel, *Linear numerical analysis* (Academic Press, 1970).

R. T. Gregory and D. L. Karney, *A collection of matrices for testing computational algorithms* (Wiley, 1969).

R. W. Hamming, *Introduction to applied numerical analysis* (McGraw-Hill, 1971).

A. S. Householder, *Matrices in numerical analysis* (Dover, 1975).

A. S. Householder, *Lectures on numerical algebra* (Math. Assoc. of America, 1972).

E. Isaacson and H. B. Keller, *Analysis of numerical methods* (Wiley, 1966).

P. Lancaster, *Theory of matrices* (Academic Press, 1969).

M. Marcus, *Basic theorems in matrix theory* (U. S. Government Printing Office, 1960).

M. Marcus and H. Minc, *A survey of matrix theory and matrix inequalities* (Allyn and Bacon, 1964).

Modern Computing Methods (H. M. Stationery Office, 1961).

B. Noble, *Applied linear algebra* (Prentice Hall, 1968).

J. M. Ortega, *Numerical analysis, a second course* (Academic Press, 1972).

E. Stiefel, tr. W. C. and C. J. Rheinboldt, *An introduction to numerical mathematics* (Academic Press, 1963).

J. Stoer, *Einführung in die numerische Mathematik*, I (Springer, 1972).

J. STOER and R. BULIRSCH, *Einführung in die numerische Mathematik*, II (Springer, 1973).
JOHN TODD, ed., *Survey of numerical analysis* (McGraw-Hill, 1962).
JOHN TODD, Chapter 7, Part I of E. U. Condon-H. Odishaw, *Handbook of Physics*, 2nd ed. (McGraw-Hill, 1967).
R. S. VARGA, *Matrix iterative analysis* (Prentice Hall, 1962).
B. WENDROFF, *Theoretical numerical analysis* (Academic Press, 1966).
J. H. WILKINSON, *The algebraic eigenvalue problem* (Clarendon Press, 1965).
J. H. WILKINSON, *Rounding errors in algebraic processes* (Prentice Hall, 1963).
J. H. WILKINSON and C. REINSCH, *Linear algebra* (Springer, 1971).
D. M. YOUNG, *Iterative solution of large linear systems* (Academic Press, 1971).
D. M. YOUNG and R. T. GREGORY, *A survey of numerical mathematics* 2 vols. (Addison-Wesley, 1972—3).

Attention is also invited to many useful expository papers in this area; some appear in Symposia Proceedings. Apart from those written by authors mentioned above, and often incorporated in their books, we mention

G. GOLUB, *Least squares, singular values and matrix approximations,* Aplikace Math. *13*, 44—51 (1968).
W. KAHAN, *Numerical linear algebra,* Canadian Math. Bull. *9*, 757—801 (1966).
O. TAUSSKY, *A recurring theorem,* Amer. Math. Monthly *56*, 672—676 (1939). *Bounds for eigenvalues of finite matrices,* pp. 279—297, in *Survey of Numerical Analysis*, ed. J. Todd (1962). *On the variation of the characteristic roots of a finite matrix*, pp. 125—138, in *Recent advances in matrix theory* (1966).
A. M. TURING, *Rounding-off errors in matrix processes,* Quart. J. Mech. Appl. Math. *1*, 287—308 (1968).
J. VON NEUMANN and H. H. GOLDSTINE, *Numerical inverting of matrices of high order,* Bull. Amer. Math. Soc. *53*, 1021—1057 (1967), Proc. Amer. Math. Soc. *2*, 188—202 (1951).

A full bibliography of Numerical Algebra has been prepared by A. S. Householder.

SUPPLEMENTARY REFERENCES

C. L. LAWSON and R. J. HANSON, *Solving least squares problems* (Prentice Hall, 1974).
A. BEN ISRAEL and T. N. E. GREVILLE, *Generalized inverses: theory and applications* (Wiley, 1974).
G. W. STEWART, *Introduction to matrix computation* (Academic Press, 1972).
A. KORGANOFF and M. PAVEL-PARVU, *Méthodes de calcul numérique*, 2 (Dunod, 1967).
A. GEWIRTZ, H. SITOMER and A. W. TUCKER, *Constructive linear algebra* (Prentice Hall, 1974).
F. H. HILDEBRAND, *Introduction to numerical analysis* (McGraw-Hill, 1974).
L. F. SHAMPINE and R. C. ALLEN, *Numerical computing* (Saunders, 1973).
H. R. SCHWARZ, H. RUTISHAUSER, E. STIEFEL, *Numerical analysis of symmetric matrices* (Prentice Hall, 1973).
R. P. BRENT, *Algorithms for minimization without derivatives* (Prentice Hall,1973).

Index

A B C 8 D 9 E 0 F 1 G 2 H 3 I 4 J 5